STRIKING A BALANCE

Improving Stewardship of Marine Areas

Committee on Marine Area Governance and Management

Marine Board
Commission on Engineering and Technical Systems
National Research Council

NATIONAL ACADEMY PRESS
Washington, D.C.   1997

NATIONAL ACADEMY PRESS • 2101 Constitution Avenue, N.W. • Washington, D.C. 20418

NOTICE: The project that is the subject of this report was approved by the Governing Board of the National Research Council, whose members are drawn from the councils of the National Academy of Sciences, the National Academy of Engineering, and the Institute of Medicine. The members of the committee responsible for the report were chosen for their special competences and with regard for appropriate balance.

This report has been reviewed by a group other than the authors according to procedures approved by a Report Review Committee consisting of members of the National Academy of Sciences, the National Academy of Engineering, and the Institute of Medicine.

This study was supported by Cooperative Agreement No. DTMA91-94-G-00003 between the Maritime Administration of the U.S. Department of Transportation and the National Academy of Sciences; and Grant No. N00014-95-1-1205 between the Department of the Navy, Office of Naval Research, and the National Academy of Sciences. Any opinions, findings, conclusions, or recommendations expressed in this publication are those of the author(s) and do not necessarily reflect the views of the organizations or agencies that provided support for the project.

International Standard Book Number 0-309-06369-8
Library of Congress Catalog Card Number 97-45293

Copies of this report are available for sale from:
National Academy Press
2101 Constitution Avenue, N.W.
Washington, DC 20055
(800) 624-6242 or (202) 334-3313 (in the Washington metropolitan area)
Internet, http://www.nap.edu

iv

Preface

ORIGIN AND BACKGROUND OF THE STUDY

The Marine Board of the National Research Council (NRC) has conducted major assessments of the scientific and technical prerequisites for exploring and understanding the nation's coastal and marine regions in a series of reports (NRC, 1989, 1990a, 1991, 1992a; Marine Board, 1993). These and other assessments have shown that the nation's interest in the conservation and wise management of ocean territory requires sustained public investment in information gathering and management. The findings of the Marine Board studies have revealed a strong interest in the nation's coastal and marine areas by present and potential offshore industries, coastal states responsible for resource development and environmental preservation of their offshore regions, and the ocean research community. Little has been done however, to devise a comprehensive regulatory or management framework for current or future activities in federal and state waters or on or under the seabed in the U.S. Exclusive Economic Zone.[1] The need for a regulatory and management framework is likely to increase in the future as advances in offshore technology and changes in market conditions lead to an increase in coastal populations and marine recreation and tourism, activities that utilize or impinge on coastal and marine resources (such as marine aquaculture and offshore oil, gas, and minerals exploration), and waste disposal in deeper waters. These activities may conflict with plans for setting aside areas as marine sanctuaries and raise concerns about ocean pollution.

---

[1]The exclusive economic zone (EEZ) is the area 200 nautical miles from each nation's continental boundary; authority over resources in this region is ascribed to the nation.

In April 1993, the Marine Board sponsored a forum at which representatives from private industry, public agencies, public interest groups, and the academic ocean policy community were invited to air their views on a national strategy to manage the nation's coastal and ocean resources and space. Based on the proceedings of the forum, the Marine Board identified emerging issues in marine area governance and management.

Many participants in the forum expressed the view that defining national goals and plans for the ocean is a critical prerequisite to appropriate economic investment and sound environmental stewardship of the ocean. A comprehensive national strategy that establishes a predictable legal and regulatory regime for ocean utilization would allay fears of rampant and destructive development on the one hand and environmental gridlock over future development on the other. A national strategy would also create a climate for a reasonable, nonadversarial approach to resolving conflicts.

A national strategy should define objectives for ocean utilization and preservation, establish governance mechanisms for the allocation of ocean resources and space, and institute a process for reconciling differences among stakeholders. This process needs to be dynamic and flexible to accommodate changing conditions. The national strategy should be developed through a full partnership among federal, state, and local agencies, beginning with the definition of principles for governance and continuing through the implementation of management processes. The strategy must take into account regional differences and concerns as well as national goals for the ocean and marine areas.

Following the forum, a planning meeting was held in July 1994, which was attended by representatives of interested and active parties in ocean governance and management. Based on presentations and discussions at the planning meeting, together with subsequent comments by the attendees, a core group of participants prepared a background paper identifying issues that needed to be addressed. The background paper was used as the basis for discussions with representatives of the responsible federal agencies to develop guidelines for improving coastal and marine governance and management (See Appendix B). The National Oceanic and Atmospheric Administration, the Minerals Management Service, and the Environmental Protection Agency provided funds for this study.

COMMITTEE COMPOSITION AND SCOPE OF THE STUDY

The NRC's Commission on Engineering and Technical Systems, under the auspices of the Marine Board, assembled a committee of 15 members to design and recommend ways to improve the management and governance of the nation's marine areas and resources. Committee biographies are found in Appendix A. Committee members had expertise in ocean resources management, marine environmental science, economics, law, and political science. Representatives were also appointed from communities of users and/or developers of ocean and coastal resources and space, including coastal state government agencies, fisheries, the marine transportation sector, offshore energy industries, and marine environmental protection organizations. The committee was subject to the usual NRC bias and conflict of interest procedures. Formation of the committee was coordinated with the NRC's Ocean Studies Board, the Board on Environmental Studies and Toxicology, and the Board on Biology. Sponsoring federal agencies and other agencies involved in marine area governance and management were asked to designate liaisons to the committee.

The committee's task included the following objectives:

- to collect information on and review the governance and management elements of marine management areas in the United States
- to assess the issues and elements of marine governance and management laid out in the conceptual framework paper by evaluating case studies of marine management areas
- to distill the lessons learned from the case studies and from the other avenues of committee inquiry into alternative models for the governance and management of marine areas

The study focused on the marine environment, which is defined here as the area between high water and the seaward extent of the U.S. Exclusive Economic Zone. Distinctions between state and federal waters were not a focus of this study. Although the committee recognizes that activities on land and in enclosed coastal waters (e.g., estuaries and bays) affect conditions in coastal and marine areas, the report does not address land-based management and governance problems.

HOW THE STUDY WAS CONDUCTED

The background paper from the planning meeting (see Appendix B) served as the conceptual framework for the committee's examination of real-world examples of current marine area management and provided preliminary criteria for improving ocean governance. Three in-depth case studies were conducted under the auspices of the committee as examples of existing marine governance and management programs and processes:

1. Gulf of Maine/Massachusetts Bay (conducted by the Marine Policy Center of the Woods Hole Oceanographic Institution)

2. Florida Keys National Marine Sanctuary/Florida Bay Ecosystem (conducted by the Center for the Economy and the Environment of the National Academy of Public Administration)

3. Southern California Coast, offshore and coastal region from the San Luis Obispo/Monterey County line in the north to the United States/Mexico border in the south (conducted by the Center for the Economy and the Environment of the National Academy of Public Administration)

Additional programs were also reviewed during the course of the study, including the Coastal Zone Management Program, the National Estuary Program, fisheries management under the Fishery Management and Conservation Act, the Outer Continental Shelf oil and gas leasing program, and state ocean and coastal management and governance programs.

After completing the case studies and other investigations, the committee distilled the lessons learned and developed the recommendations presented in this report. The report is available to the general public but is designed for use by the following audiences: federal and state agencies with direct management or regulatory responsibilities for coastal and marine areas or particular resources; scientists and policy analysts with a particular interest in marine and coastal species, habitats, and ecosystems and associated economic, social, and political problems; individuals, organizations, and companies with an interest in the utilization or preservation of marine and coastal resources and the services they provide; international organizations attempting to establish marine reserves and parks or to develop marine area management structures; environmental and resource public interest groups and the interested public; and key congressional staff.

The case studies are examples of diverse management requirements and geographic regions. In each case study, the committee examined an activity or issue of local importance to reveal the success or failure of existing management and governance structures and processes. Although other activities or concerns might also have been of interest to this study, the committee had only enough time and resources to examine the major issues in each case study area. The Florida Keys case study focuses on the national marine sanctuary, where concerns about ecosystem preservation conflict with intensive recreational use and resource development. The Gulf of Maine case study examines an ocean resource of substantial economic value (commercial fishing) in a state of crisis. The Southern California coast is intensely used for various activities, including marine transportation and recreation; controversy surrounding the development of offshore oil and gas resources has sparked ongoing conflicts between local and federal interests.

Each case study area was assessed with regard to (1) the effects of local, state, and federal regulations; (2) the ecological and biological issues of concern; (3) the potential for commercial or recreational uses; and (4) the social, cultural, and economic context. Criteria to guide the conduct of the case studies were based on the analyses in the background paper (see Appendix B) and the deliberations of the committee. Regional perspectives and expertise were sought through meetings in the case study areas (see Appendix C). Representatives of federal agencies responsible for marine management in these areas provided information and avenues for the exchange of information and also shared their expertise.

The case studies were performed under the guidance of individual committee members or subgroups, as appropriate. Based on lessons learned from the case studies and other activities, the committee developed recommendations for improving the governance and management of marine areas both for environmental stewardship and for the development of ocean resources.

REPORT ORGANIZATION

Chapter 1 provides an overview of the importance of marine areas and resources to the nation and an expanded definition of the terms, scope, and approach of the project, as well as a vision for the future and principles to guide the committee's analysis. Chapter 2 discusses the value of the marine environment in economic and other terms, assesses outstanding problems, and describes existing management and governance institutions and processes. Chapter 3 presents the committee's analysis of lessons learned from the case studies and examinations of other programs. Based on this analysis, Chapter 4 provides criteria for evaluating the effectiveness of management and governance. Chapter 5 proposes a number of options for improving marine governance. Chapter 6 presents tools for improving the management of marine areas. Chapter 7 contains the committee's conclusions and recommendations. The committee's findings, conclusions, and recommendations are also summarized in the Executive Summary. The case studies and a paper examining trends in coastal conditions are available upon request from the Marine Board (Bacon, 1996). Appendices provide information on committee members (Appendix A), the background paper for the study (Appendix B), a list of participants and contributors (Appendix C), and an expanded discussion of options for financing governance and management (Appendix D).

ACKNOWLEDGMENTS

The committee wishes to thank the many individuals who contributed their time and energy to this project through presentations, correspondence, and discussions with committee members. Particular appreciation is extended to the liaison representatives of the sponsoring and other interested federal agencies, who participated in meetings and provided invaluable assistance in gathering information about existing programs: Jeffrey R. Benoit of the Office of Ocean and Coastal Resource Management at the National Oceanic and Atmospheric Administration (NOAA); Thomas E. Bigford and William W. Fox, Jr., of the National Marine Fisheries Service, NOAA; Darrell D. Brown of the Office of Wetlands, Oceans, and Watersheds at the Environmental Protection Agency; Paul Stang of the Minerals Management Service; Raul Pedrozo of the U.S. Navy; and Lou Orsini and Wesley Marquardt of the U.S. Coast Guard. Special appreciation is extended to the directors and staff of the Rosensteil School of Marine and Atmospheric Sciences, in Miami, Florida, the New England Aquarium in Boston, Massachusetts, and the Port of Oakland, California, for hosting committee meetings.

The National Academy of Sciences is a private, nonprofit, self-perpetuating society of distinguished scholars engaged in scientific and engineering research, dedicated to the furtherance of science and technology and to their use for the general welfare. Upon the authority of the charter granted to it by the Congress in 1863, the Academy has a mandate that requires it to advise the federal government on scientific and technical matters. Dr. Bruce M. Alberts is president of the National Academy of Sciences.

The National Academy of Engineering was established in 1964, under the charter of the National Academy of Sciences, as a parallel organization of outstanding engineers. It is autonomous in its administration and in the selection of its members, sharing with the National Academy of Sciences the responsibility for advising the federal government. The National Academy of Engineering also sponsors engineering programs aimed at meeting national needs, encourages education and research, and recognizes the superior achievements of engineers. Dr. William A. Wulf is president of the National Academy of Engineering.

The Institute of Medicine was established in 1970 by the National Academy of Sciences to secure the services of eminent members of appropriate professions in the examination of policy matters pertaining to the health of the public. The Institute acts under the responsibility given to the National Academy of Sciences by its congressional charter to be an adviser to the federal government and, upon its own initiative, to identify issues of medical care, research, and education. Dr. Kenneth I. Shine is president of the Institute of Medicine.

The National Research Council was organized by the National Academy of Sciences in 1916 to associate the broad community of science and technology with the Academy's purposes of furthering knowledge and advising the federal government. Functioning in accordance with general policies determined by the Academy, the Council has become the principal operating agency of both the National Academy of Sciences and the National Academy of Engineering in providing services to the government, the public, and the scientific and engineering communities. The Council is administered jointly by both Academies and the Institute of Medicine. Dr. Bruce M. Alberts and Dr. William A. Wulf are chairman and vice chairman, respectively, of the National Research Council.

Contents

xi

List of Boxes

Executive Summary

Developing a coherent framework to guide the nation's activities in the ocean and coastal regions is especially important in this time of growing national interest in the ocean, which includes heightened awareness of the need to protect it, along with recognition of new opportunities to utilize marine resources. Such a framework is necessary to guide the nation's activities in the ocean and coastal regions. Challenges to the current system have arisen from changes in national priorities and in the international economic system, including the recognition that good environmental policies make good economic policies, the challenges of the globalization of markets and opportunities, and a new willingness for the U.S. government to become a catalyst for technology development and economic growth, as well as a steward of the nation's natural resources.

At the same time, demands on the coastal marine environment have been intensifying through the rapid migration of people to the coasts, the growing importance of the coasts and ocean as areas for aesthetic enjoyment, and increasing pressures to develop ocean resources and spaces for economic benefits (e.g., commercial fisheries, marine aquaculture, marine energy, and mineral resources). Taken together, all of these factors have created a sense of urgency about developing a coherent national system for making decisions.

The overall value of a healthy, diverse, and productive marine environment is difficult to assess in quantitative terms but is indisputably immense. An improved system of marine area governance and management will be effective only if it is perceived as helping to meet the national interests in the marine environment. The national interest is defined here not as the interest of the federal government. It denotes instead the fundamental values the nation as a whole has embraced for the protection and use of the marine environment. This definition transcends the interests of any single agency, mission, or special interest group and presupposes a reasonable accommodation of the expectations of competing interests, as well as protection of the basic fabric of the functioning marine environment.

The process of marine area governance has two dimensions: a political dimension (governance), where ultimate authority and accountability for action resides, both within and among formal and informal mechanisms; and an analytical, active dimension (management), where analysis of problems leads to action. In practice, there is a continuum from governance to management. The present governance and management of our coastal waters are inefficient and wasteful of both natural and economic resources. The primary problem with the existing system is the confusing array of laws, regulations, and practices at the federal, state, and local levels. The mandates of various agencies that implement and enforce existing systems often conflict with each other. In many cases, federal policies and actions are controlled from Washington with little understanding of local conditions and needs. No mechanism exists for establishing a common vision and a common set of objectives.

Managing marine resources presents special challenges: marine resources are in the public domain, so the incentives provided by private property rights and market signals are largely absent. Many marine resources and resource users are mobile, creating ample potential for interference and conflicts; users often operate offshore, where monitoring and enforcement of rules is difficult. For these and other reasons, effective governance would be difficult at best, but the difficulties are compounded by the fractured framework of laws, regulations, and practices at the federal, state, and local levels.

As the intensity of use of the marine environment grows, the lack of effective governance is rapidly becoming a critical problem. The biological integrity of the sea is being steadily impaired, as has been demonstrated by declining fish resources and the loss of critical coastal habitats. In addition, growing conflicts about, and intensity of use of, marine resources often result in wasted economic or social opportunities. These problems will inevitably become more acute as growing populations, which are increasingly concentrated on the coast, continue to put stress on this critical global resource.

Many organizations and groups are involved in governing and managing resources and activities in marine and coastal areas, including federal, state, and local governmental agencies; commercial and industrial interests; recreational users; and environmental groups. Each group typically has a direct interest in governance and management and seldom coordinates activities with other organizations operating in the area.

These conditions were apparent in the case studies of marine area governance and management in southern California, the Florida Keys, and the Gulf of Maine, as well as in other federal and state marine management activities examined by the committee. In addition to focusing on particular problems in each area, the case studies identify efforts to improve governance and management. In southern California, a collaborative effort among local, state, and federal agencies has forged a consensus among the interests for a plan to develop offshore oil resources. In the Florida Keys, local officials of the National Marine Sanctuary program are building support for the marine sanctuary planning process. In the Gulf of Maine, the Gulf of Maine Council is developing common goals and objectives among the states and provinces that border this body of water to address regional economic and environmental issues. These efforts have had some success developing plans that consider marine areas holistically. More important, all of them involve broad cross-sections of stakeholders at the local level.

None of the initiatives described above originated in Washington, D.C.; they were carried out by officials and stakeholders familiar with local problems who believed they could find solutions by working together. In many cases, efforts were initiated by a single individual who had the conviction and courage to go beyond the norms of bureaucratic behavior and try a new approach. No mechanism exists today for nurturing this type of initiative or for ensuring its continuity when the key individuals are no longer directly involved. Existing government mechanisms typically operate through relatively rigid hierarchical structures.

As a result of the case studies and other investigations of existing marine and coastal programs, and based on the performance standards developed at the outset of this study, the committee concluded that any system for improving the governance and management of the coastal areas must include the following elements:

- a method for developing common goals and objectives in harmony with broad national interests
- opportunities for policy-making and decision-making authority at the local or regional level
- effective management tools designed to deal with the particular problems of resource use in the marine environment

Significant large-scale changes to existing systems of governance and management will be required before improvements can be realized. These changes will also require substantial, sustained efforts on the part of the organizations involved.

A number of precedents for successful, large-scale organizational change have been established in recent years. Many corporations, government agencies, military organizations, and volunteer groups have redesigned their approaches to management in the face of rapid changes. The techniques, tools, and experiences of these organizations have been documented and can be used as guidelines for redesigning marine and coastal governance and management systems. Attempts to implement new systems without fundamentally changing the way things are done today are likely to fail (National Performance Review, 1993).

The new design must consider all aspects of the existing system, including roles and responsibilities, authorities, relationships among departments and agencies and levels of government, information systems and databases, and recognition and reward systems. Changes might not be required in all of these elements, but care must be taken to ensure that they are all compatible with the new system.

FEDERALIST APPROACH

In addition to the governance problems created by multiple nonmarket uses of marine resources and maintaining access to them, existing systems have two fundamental problems—first, fragmentation among federal and local agencies and second, not enough participation and coordination of interests at the local level. The findings of this study indicate that these problems can best be addressed by adopting a federalist form of governance modeled after the distribution of power between the federal government and the states. In this instance, however, federalism is not about a separation of power between federal and state governments. Instead, a federalist system of governance places power at the appropriate level for accomplishing objectives and implementing actions. A federalist approach would lead to better protection and promotion of the national interest in the long-term health and efficient use of the marine environment, while being responsive to, and building on the capacity of, local and regional interests.

One of the main tenets of federalism is that authority belongs at the lowest point in the organization that has the capability and information to get the job done. The top level of the organization establishes the broad framework and ground rules under which the organization operates. It is responsible for defining the purpose, values, and vision of the organization and for establishing expectations and a system for measuring outcomes. Within this overall framework, the local group, which could include representatives of federal, state, and local governments and other stakeholders, assumes the responsibility and authority for charting and managing its own course of action. The local group is, however, accountable to the top level of the organization and must provide ample and timely feedback.

In a federalist structure, the top level of the organization serves an ongoing role as the enabler of the process by creating an environment that allows local groups to make their own decisions by providing training, by offering advice when requested, by serving as a repository of technical expertise, and by supporting the implementation of actions after decisions have been made. The top level of the organization also provides mechanisms for reconciling differences among decentralized authorities. Federalism recognizes that each area is unique, that each local group faces unique problems and must develop strategies and plans to handle them. The following recommendations are intended to provide a framework for improving the nation's stewardship of valuable and irreplaceable marine resources.

CONCLUSIONS AND RECOMMENDATIONS

Defining basic principles and effective processes for improved governance of ocean and coastal areas is a prerequisite both to sound economic investment and environmental stewardship and creates a climate for a reasonable, less adversarial approach to resolving conflicts. General elements of the framework for improved governance and management envisioned in this report include the following:

- There must be a clear statement of goals, especially where different entities must be brought together in a cooperative management effort.
- The geographic (or ecological) area to be managed needs to be carefully delineated.
- Mechanisms need to be designed for involving all relevant stakeholders in the governance process.
- In most situations, the process should be initiated as a joint state-federal effort.
- Systems should foster innovative responses to management needs and opportunities for resource utilization.
- Processes should facilitate the incorporation of scientific information into all aspects of decision making.
- Success should be measured by a clear system of monitoring and evaluation.

The system recommended in this report has four basic components:

- creation of a National Marine Council to improve coordination among federal agencies, monitor the marine environment, facilitate regional solutions to marine problems, and facilitate interagency problem solving
- creation of regional marine councils where they are needed to provide innovative approaches to complex marine governance issues at the operational level
- enhancement of the ability of individual federal programs to succeed in their missions
- adoption of management tools that would increase the effectiveness of regional councils and individual agencies

National Marine Council

The National Marine Council would be made up of directors of federal ocean and coastal agencies and would report directly to the President. The council would develop goals, principles, and policies for resolving issues of marine governance; review existing federal legislation; and coordinate national goals by balancing environmental protection with appropriate development of resources. The council would also oversee efforts to address other relevant national concerns, such as the protection of human health and safety and national security in relation to marine resources and areas. Other functions of the National Marine Council would include surveillance of the marine environment, identification of marine area problems and conflicts, and encouraging innovative ways to resolve regional problems. The National Marine Council would ensure that the United States has clearly identified global marine issues and has mobilized adequate resources to address them.

Regional Marine Councils

In situations where there are long-standing conflicts among local or regional interests or where there are risks to marine resources or the environment, the National Marine Council should encourage the formation of regional councils. Regional councils would provide technical assistance on marine management issues, ensure the application of scientific and monitoring information, develop alternative processes for resolving disputes, encourage participation by local interests in governance decisions, and pursue contractual arrangements with stakeholders and other participants.

Regional councils would only be used in high value, high conflict, high risk, or high damage areas. They would remain in existence only for the duration of the problem or conflict but would not be permanent bodies. The composition of each regional council would vary according to the problem and the region. Functions of the regional councils would include developing long-range goals for the region and plans for achieving those goals, coordinating planning and management among state and federal agencies, coordinating fiscal planning for pooling regional resources, mediating disputes among agencies and stakeholders, and executing contracts with various groups to resolve and manage specific problems.

Improving Existing Programs

Existing federal and state coastal and marine management programs could become platforms upon which to base improved governance and management structures and processes. Recommendations for improving some existing programs are found in Chapter 5 of this report. Generally, however, all existing programs could become more effective by coordinating their activities with other federal, state, and local agencies that have jurisdictional or management responsibilities, by involving stakeholders and nongovernmental groups in decision making, and by adopting area-based views that take into account regional ecology, the array and condition of resources, and by balancing environmental and economic considerations. Existing programs would also benefit from a broader range of management tools for dealing with problems and conflicts. Federal officials, in cooperation with their state counterparts, should maximize existing programs, especially where there are urgent problems. Most existing programs could be reconfigured to deliver some, or all, of the elements associated with regional councils.

Improving Management Tools

Institutions charged with designing and applying policies have a variety of management tools with which to address problems associated with the use of marine resources and space. No single instrument is appropriate under all circumstances. Selecting a management tool involves weighing historical, technical, and economic factors, as well as social and political factors. Many innovative management tools have been used, on a limited basis, in the marine context or in the terrestrial environment. These tools include zoning and the creation of refuges, systems for establishing liability for environmental or other damage, compensation for the economic losses of certain stakeholders, user charges and transferable entitlements to regulate demands on marine resources, and negotiating ways to mitigate activities that harm marine resources or space. These management tools are discussed in detail in Chapter 6 of this report. These tools should be given renewed attention and should be more widely used in existing marine management programs and in the proposed regional marine councils. Recommendations for expanding the use of these tools are given in Chapter 7.

A fully developed system that meets all of the objectives and contains all of the elements discussed in this report must necessarily evolve over time in response to actual experience. However, the committee believes opportunities are available for moving forward now by improving existing marine management programs.

REFERENCE

National Performance Review. 1993. From Red Tape to Results: Creating a Government that Works Better and Costs Less. Washington, D.C.: U.S. Government Printing Office.

1

Introduction

BACKGROUND AND OVERVIEW

Growing national interest in the ocean, awareness of threats to marine integrity, and opportunities to utilize marine resources make this an ideal time to explore a more coherent system of governance to guide activities in the ocean and coastal regions. Challenges to existing systems have arisen from changes both in national priorities and in the international economic system, including the recognition that good environmental policies make economic sense, the globalization of markets and opportunities, and a willingness on the part of the government to become a catalyst for technology development and economic growth as well as a steward of natural resources.

At the same time, demands on the coastal marine environment are intensifying through the continued migration of people to the coasts, the growing importance of the coasts and ocean for aesthetic enjoyment, and increasing pressures to develop ocean resources and space for economic benefits (e.g., commercial fisheries, transportation, marine aquaculture, marine energy, and mineral resources). The combination of these factors has created a sense of urgency for the development of a coherent system for making decisions in this arena (NRC, 1995c).

Although the marine environment is typically open to multiple uses, the United States now manages ocean and coastal space and resources primarily on a sector-by-sector basis. For example, to a large degree, one law, one agency, and one set of regulations govern offshore oil and gas; a different law, agency, and regulations govern fisheries; and other single-purpose regimes oversee water quality, navigation, protected areas, endangered species, and marine mammals (Cicin-Sain and Knecht, 1985). Although regimes for the management of resources are established on a statute-by-statute basis, each area of interest may, at the same time, be impinged upon by a plethora of other regulatory management regimes. Except for the modest, but important, marine sanctuaries program and a few emerging state programs, ocean regions are not managed on an area-wide, multi-purpose, or ecological basis; nor are there agreed upon processes for making trade-offs and resolving conflicts among various interests (NRC, 1995c).

The single-purpose approach to management often leads to adverse ecological impacts, economic stagnation, and political gridlock. Single-purpose laws do not account for the effects of one resource or use on other resources or on the environment as a whole. They do not assess cumulative impacts, and, therefore, may not provide a basis for resolving conflicts. Even if a management regime encompasses both economic and environmental considerations, implementation in a specific situation has frequently meant not accommodating the concerns of conflicting interest groups. In the absence of an integrated framework, parties seeking to utilize ocean resources and space for economic objectives and parties concerned with environmental preservation often have to rely on the exercise of political power or litigation as primary mechanisms for achieving their aims. Significant societal and economic costs are incurred through these adversarial processes (Outer Continental Shelf Policy Committee, 1993).

Previous Marine Board studies of issues associated with the nation's ocean space and resources (NRC, 1989, 1990a, 1991, 1992a; Marine Board, 1993) have confirmed that the absence of a coherent national system of governance for marine

Container ship sailing into the Port of San Francisco. Photo courtesy of the Marine Board.

Searun, Inc., Salmon Farm, Eastport, Maine. Photo courtesy of Katharine Wellman.

resources and the use of ocean space have often led to economic stagnation and political stalemate. These reports have also concluded that a more coherent process for governing marine activities and resources would ensure that the nation's ocean ecosystems and their living resources were protected. At the same time, economic development would be allowed, where appropriate. In a coherent system, existing and potential conflicts among different users of the ocean would be anticipated and addressed through mechanisms for the equitable allocation of ocean space and resources in keeping with national stewardship over the region.

The general basis for this approach is articulated in a report by the President's Council on Sustainable Development (President's Council on Sustainable Development, 1996), in which a national commitment is made to "a life-sustaining Earth....A sustainable United States will have a growing economy...and...will protect its environment, its natural resource base, and the functions and viability of natural systems on which all life depends."

CHALLENGE OF MANAGING MARINE AREAS

Marine areas extending from the coastline of the United States to 200 miles offshore are immensely valuable resources to the people of the United States. They provide:

- habitat for a wide range of plant and animal species that are essential to the global ecosystem
- fish and shellfish that support the majority of commercial and recreational fisheries
- reserves of oil, gas, and other minerals
- travel-ways for coastal and international shipping and the maneuvering area for the U.S. Navy
- places for swimming, boating and other outdoor recreational activities that provide renewal and relief from the pressures of everyday life
- a basis for the tourism and recreation industries
- access to coastal development
- important influences on the climate of coastal regions
- essential aspects of our culture, traditions, and heritage

The marine areas that provide these benefits have always been publicly owned and open to all users. Within three miles of the shore, underwater lands are in the domain of the states. From three miles to 200 miles offshore, they are overseen by the federal government on behalf of all Americans. Sustaining the ecological health and economic productivity of this vast underwater commons requires careful, informed, effective, and decisive management.

Unfortunately, management practices have not kept pace with growing pressures on the marine environment. The population living in coastal areas has grown

Great blue heron, Chincoteague National Wildlife Refuge, Virginia. Photo courtesy of William Eichbaum.

rapidly (Culliton et al., 1990). Technology has extended the power of humans to exploit marine resources. The demand for products from our oceans, gulfs, and bays has increased dramatically. Individual users and interest groups have become more defensive about the benefits they obtain from coastal waters and more strident in their efforts to secure diminishing resources for themselves.

A wide range of state and federal agencies and programs have been created to respond to these pressures. In many instances, these programs have succeeded in reversing environmental decline and in resolving conflicts. But in a growing number of cases, the institutions responsible for managing marine areas have not anticipated the ecological risks of their actions or inaction, have not been able to coordinate and sustain efforts to solve large-scale marine management problems, and have been paralyzed by conflicts among interest groups.

The depletion of fish stocks in the Gulf of Maine is the most notable recent example of the failure of the management of marine resources. The historic, and ongoing, decline of productive natural systems like the Chesapeake, San Francisco, and Florida bays is further evidence of the need to change our approach to marine area management (Fogerty et al., 1991; Hedgepeth, 1993). Global climate changes, continuing population growth, advances in mechanical, electronic, and biological technology, the expansion of international trade, and incursions of nonindigenous species all suggest a need for more effective management of marine areas (NRC, 1995c).

DEFINITION AND SCOPE OF THIS STUDY

The process of marine area governance has two dimensions: a political dimension (governance), where ultimate authority and accountability for action reside, both within and among formal and informal mechanisms; and an analytical, active dimension (management), where analysis of problems leads to action. In practice, there is a continuum from governance to management. The country needs a coherent system of governance, based on a set of overarching principles and processes, and an appropriate set of management tools vigorously applied to deal with the unique characteristics of marine resources. A great many tools presently deployed in the marine and coastal environment embody one form of management or another. But there is no coherent system of governance.

This study attempts to identify principles and goals, as well as elements, processes, and structures for improving marine area governance. The concepts outlined below are applicable in the marine environment, that is the area between the high water line and the seaward extent of the U.S. Exclusive Economic Zone (EEZ).[1] The geographic area of concern for this study is the marine environment

---

[1]The exclusive economic zone (EEZ) is an area 200 nautical miles beyond and adjacent to the territorial sea, under which the coastal state has sovereign resource rights and jurisdiction with regard to marine scientific research and protection and preservation of the marine environment.

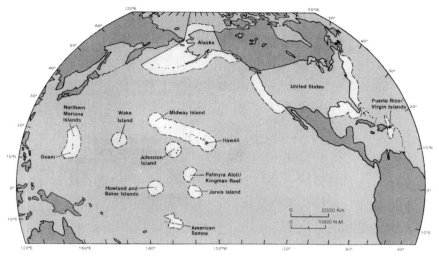

U.S. Exclusive Economic Zone.

The Exclusive Economic Zone of the United States and its Trust Territories (shaded areas). Illustration courtesy of the National Ocean Service, the National Oceanic and Atmospheric Administration, U.S. Department of Commerce.

of the United States, including coastal resources such as bays and estuaries, without regard to current determinations of jurisdictional authority between the states and the federal government.

The term "marine management area" used in this study refers to an area for which coherent plans are developed and measures taken to govern the uses of the area systematically. Types of marine management areas include sanctuaries, reserves, parks, and other units subject to regional planning and management. Planning by states to manage their ocean and coastal areas also represent efforts to implement this concept. Certain federal statutes, such as the Coastal Zone Management Act (CZMA), the Fisheries Conservation and Management Act and the Outer Continental Shelf (OCS) Lands Act, have also established processes with some of the characteristics of marine area management.

Until recently, prohibiting or limiting activities in specific ocean areas has been the primary strategy for controlling development. This approach neither mediates differences among multiple users nor ensures ecological integrity. An integrated system of governance is necessary for managing these regions and resources in keeping with the best interests of the nation—for present and future generations.

A number of federal and state agencies have jurisdiction over marine and coastal areas, activities, and resources. The National Oceanic and Atmospheric Administration (NOAA) is responsible for the marine sanctuary program; for fisheries management; and for providing states with a national framework for coastal management, including funding grants for ocean management planning. The U.S. Environmental Protection Agency (EPA) is responsible for maintaining the quality of the nation's waters and has several regional planning programs that directly address the uses and management of ocean regions, such as the Gulf of Mexico and the National Estuary Program (NEP). EPA is also responsible for designating and managing ocean disposal sites. The Minerals Management Service (MMS) is responsible for the management of outer continental shelf energy and mineral resources. The National Park Service has a number of ocean properties and management responsibilities, for example, in the Channel Islands, off California, and in the Florida Keys. The U.S. Coast Guard enforces laws governing fisheries and other laws pertaining to the oceans. The U.S. Department of State is responsible for the international implications of the regional management of marine resources and uses. The U.S. Department of Defense has a keen interest in marine area governance because of the relationship of ocean space management to national defense. States, international agencies, and other countries also have responsibilities and interests.

Some coastal states have developed comprehensive plans and policies for regulating or encouraging activities that affect their territorial waters and/or coasts, but the scope of their activities is limited by their state jurisdictions. The primary goal of these management plans is to ensure sustained, long-term benefits from coastal space and resources for present and future generations. States could be substantially assisted, however, by leadership from the federal government in dealing with regions that are beyond their legal purview but where they have strong interests.

VISION OF THE FUTURE

Defining basic principles and effective processes for the governance of ocean and coastal areas is a prerequisite to both sound economic investment and effective environmental stewardship and makes it possible to establish a reasonable, nonadversarial approach to resolving conflicts. A critical examination and assessment of current practices is the basis for developing a structure and processes for the future.

The Committee on Marine Area Management and Governance envisions an improved system of governance of the marine resources of the United States that would:

- restore the ecological health and character of marine areas and sustain them permanently for the benefit of future generations
- perceive risks to the marine environment early enough to take cost-effective measures to avoid both the irrevocable loss of ocean resources and unproductive conflicts among competing interests
- enable citizens to enjoy the benefits of marine areas and resources

Principles for Marine Area Governance and Management

The committee first reviewed the background paper that emerged from the planning session for this project (Appendix B) and distilled several principles or criteria from it that define successful governance. Many of these principles reflect the 16-point "belief statement" that underpins the report of the President's Council on Sustainable Development (President's Council on Sustainable Development, 1996). The committee tested these principles in evaluating three major case studies and other examples of management described by participants at meetings in the case study regions. The modified principles became the following performance standards for successful marine area governance:

- *Sustainability*. Sustainable use of the marine environment and resources requires that the needs of the present generation not compromise the needs of future generations.
- *Regional ecosystem perspective*. Governance systems should be based on an understanding of the natural ecosystem. Strict adherence to political or jurisdictional boundaries can hamper effective governance where events, issues, and natural processes cross jurisdictional boundaries. The ecological region should include the adjacent terrestrial systems, as necessary to ensure effective resource management and governance.
- *Global imperative*. Although good regional governance is essential for good management, global issues, such as global climate change, require major policy direction on the national and international level. The decision to address these critical issues cannot be made at the regional level although innovative measures for addressing them may be developed and implemented there.
- *Adaptive management*. The system should be able to accommodate changes in scientific understanding and advances in technology and to recognize that social values can shift the fundamental requirements and constraints of governance. Management should be viewed as a learning experience for approaching future problems.
- *Scientific validity, including risk assessment*. Governance decisions and decision-making processes should be based on biological, physical, chemical, and ecological information, as well as cultural and social norms. Governance should include an assessment of the potential risks of action and inaction.

- *Conflict resolution.* The governance system should provide mechanisms for resolving conflicts that are fair and that reduce the delays associated with disputes.
- *Creativity and innovation.* Governance systems should foster creativity and innovation by government officials and other affected parties. This means that measured risk-taking should be rewarded and that new approaches to old problems should not be rejected because of existing regulations or government structures.
- *Economic efficiency.* The goal of governance and management should be to increase the total social value derived from marine resources. This requires giving appropriate weight to nonmarket resource values and services as well as commercial values.
- *Equity and transparency.* The governing process and decisions for allocating benefits and costs should conform to accepted norms of equity. The governing process should establish a level playing field for competing stakeholders and users, provided that equity also extends to future generations. Transparency refers to the principle that everyone affected should understand how and under what conditions they can participate in the decision-making process.
- *Integrated decision-making.* Governance structures should bring together the concerns of various agencies and stakeholders to encourage decisions that address ecological, social, economic, and political problems.
- *Timeliness.* Governance systems should operate with sufficient speed to address threats before they become crises and to meet schedules mutually agreed upon by the participants.
- *Accountability.* Authorities and structures for governance and management need to be clearly defined so that it is clear who is responsible for particular tasks and who must change policies or actions for the adaptive management process to be effective.

Underlying all of these criteria is the notion expressed in the report of the President's Council on Sustainable Development that "...in order to meet the needs of the present while ensuring that future generations have the same opportunities, the United States must change by moving from conflict to collaboration and adopting stewardship and individual responsibility as tenets by which to live" (President's Council on Sustainable Development, 1996).

2

The Way Things Are Now

NATURE AND VALUE OF THE MARINE ENVIRONMENT

Since the beginning of civilization, the ocean has played a critical role in the well-being of humankind as a source of food, a medium of transportation, and a site for recreation, adventure, and inspiration. As the twentieth century draws to a close, increased scientific understanding, combined with more intense utilization of the ocean's resources, have revealed that the resources of the marine environment are greatly affected by human activities. The ocean and its bounty appear to be susceptible to the adverse effects of human abuse whether through conflicting uses, overutilization of specific resources, or destructive human activities that degrade essential marine functions.

The marine environment is different from the terrestrial environment in that many of the processes and functions critical to its integrity occur over very long distances and time scales. For example, the El Niño phenomenon of the Pacific Ocean, which is crucial to both the climate of California and the productivity of its coastal waters, behaves in markedly different ways depending upon global meteorological events. Earth's oceans vary greatly in density, salinity, and temperature, which results in the development of a wide variety of life forms that are different from each other and different from life forms on land (Norse, 1993). More than 90 percent of all classes of organisms live in the oceans, and nearly half of all phyla are marine phyla (Weber and Gradwohl, 1995).

Although we have a limited understanding of marine ecology and biological diversity, it is becoming increasingly apparent that the ocean is a rich and diverse habitat for life on earth (NRC, 1995c). Life forms and other resources of the marine environment offer substantial benefits to humanity. Critical coastal

Notice of beach closure because of pollution. Photo courtesy of the Marine Board.

habitats provide spawning grounds, nursery areas, as well as shelter and food for finfish, shellfish, birds, and other wildlife. Approximately 85 percent of commercially harvested fish depend on estuaries and near coastal waters at some stage in their life cycle. Estuarine marshes provide natural buffers against floods and other episodic events.

National Interests in the Ocean

National interests in the ocean include protecting national security, facilitating domestic and international commerce, protecting and using sustainably the natural resources under public stewardship, and ensuring the health and safety of the American people. Each of these interests demands active and effective governance. Achieving a reasonable balance among these interests when they conflict, as they often do, poses an even greater challenge.

For reasons discussed throughout this report, the operational focal point for improving marine area governance for the United States appears to be at the regional level (for example, the New England region or the Gulf of Mexico region). However, broader national interests, such as the ones listed below, can best be articulated and protected at the federal level:

- fulfilling treaty obligations, such as obligations in the Law of the Sea Convention, that give other nations the freedom to use our EEZ for navigation and overflight[1]
- fulfilling bilateral agreements with adjacent nations (Canada, Mexico, and the Bahamas) regarding marine boundaries, migratory resources, and other transboundary issues
- maintaining national security through the use of offshore military operation and exercise areas, weapons and missile testing areas, and other ocean-based facilities
- implementing federal laws where a national interest has been formally declared (e.g., protecting marine mammals and endangered species or meeting and maintaining federal standards of air and water quality)
- maintaining the freedom of interstate commerce
- ensuring that living marine resources in waters under federal jurisdiction are sustainably managed to protect the interests of future generations
- ensuring that the public secures an adequate and appropriate rate of return for the private use of publicly owned ocean resources under federal jurisdiction

Value of Marine and Coastal Resources

The security, commercial, stewardship, recreational, and other interests in the ocean reflect the value of marine resources and the services they provide to society. However, because of their public nature, many of the services rendered by marine resources are not directly marketable (that is, they are not traded in conventional markets where the value of resources and services can be determined through prices and quantities traded). For example, although people pay implicitly through travel and related expenses to get to a recreational site (Freeman, 1995), recreational uses of the marine environment, such as boating, fishing, and wildlife viewing, are essentially free (although parking and shore access may not be). Consequently, the importance of marine recreation, noncommercial marine species, and ecosystem functions and services are often underestimated.

At the other extreme, some people assume that these "priceless" benefits warrant almost unlimited economic sacrifices. Unless realistic values are assigned to marine resources, it will continue to be difficult to strike an appropriate balance among competing ocean interests. But the marine environment cannot provide unlimited benefits. As coastal populations grow and incomes continue to

---

[1]Although the United States has not formally ratified the 1982 Law of the Sea Convention, the United States has traditionally maintained that the convention (except Part XI on deep seabed mining) generally confirms existing customary international law and practice and is, therefore, binding on all nations. In 1994, President Clinton forwarded the convention to the Senate for advice and consent to accession. Thus, the United States is bound to act in ways that do not contravene the spirit and intent of the convention.

Snorkeling in the Florida Keys National Marine Sanctuary. Photo courtesy of William Eichbaum.

rise, limiting demand is essential. Otherwise resources and services will inevitably be overexploited or their quality diminished.

Estimates of marine resource values can inform decision makers and the public about the true costs (including lost resource values) of commercial development as well as the true costs of restricting commercial development in the interests of conservation and preservation. Estimating the economic value of resource services is one way of including economic factors in trade-off decisions, of assessing damages for liability or compensation, and of deciding whether investments in the management or enhancement of these resources are economically justifiable (Bingham, 1995). The case studies prepared for this report and other materials indicate that the economic value of recreation and the passive use services of the U.S. marine environment loom very large, so large that they overshadow the value of commercial uses in many regions. For example, tourism and recreation are by far the most valuable uses of the marine environment in the Gulf of Maine and in the Florida Keys (John, 1996a).

The term *value* has several different meanings. For example, it can refer to a set of moral, ethical, or aesthetic judgments, or it can refer to economic worth. All of these meanings are relevant in marine governance, but most conflicts arise over economic trade-offs. The economic value of a marine resource is the value of the services it provides to people, in terms of their wants and preferences. These services may be provided directly, such as the opportunities a fish stock affords for catching fish, or indirectly, such as the support marine habitats provide as fish spawning grounds. The *economic value* of a good or service to the individual is the dollar amount a person is willing to pay for it.

Techniques for Measuring Values

The economic framework for valuing natural resources recognizes that natural resources, like capital assets, yield a flow of services (commercial, recreational, ecological, and nonuse). These services can be categorized as active use services, passive use services, or indirect use services. Active use services include well defined "direct" and observable uses, such as recreational boating and fishing, commercial fishing, and navigation. Passive use services include the value individuals place on natural resources apart from their own identified, measured active uses. Passive use services are often further separated into bequest and existence use services. Bequest use services reflect the value associated with maintaining the availability of natural resources for use by others today and by future generations. Existence use services reflect the value individuals associate with protecting a resource and "just knowing it is there," even if no future use of the resource is envisioned. Indirect use services incorporate elements of active and passive use services. Indirect use services incorporate the value individuals place on a resource based on potential future active use. Consideration of future

Pleasure sailing in Frenchman Bay, Maine. Photo courtesy of Katharine Wellman.

generations can be a component of indirect use services, although it is also integral to bequest use services as noted above. This component of indirect use services is sometimes referred to as option use services. In addition, indirect use services incorporate the values individuals associate with more passive use services, such as enjoying a site when driving, walking, or working nearby; and enjoying hearing about, reading about, or seeing photographs of a site. A perceived increase in human health risks may also fall within the category of indirect use services because it involves public perceptions regarding the risk that an event (e.g., containment failure of an in-water contaminated disposal facility) may occur and the public's willingness to pay to avoid it.

A variety of techniques can be used to estimate people's willingness to pay for (or willingness to accept) marine resources and resource services even if no direct payment for them is required. These methods include the travel cost model, random utility model, hedonic price method, and contingent valuation. All of these methods have been tested in a number of studies related specifically to marine resources and resource services, especially marine recreation (Freeman, 1995) and commercial fisheries (Lynne et al., 1981; Bell, 1972, 1989). Studies have also been conducted linking water quality with the demand for various recreational activities (Kaoru et al., 1995; Bockstael et al., 1995). Less work has been devoted to estimating the willingness to pay for marine ecosystem functions and services and for passive uses.

Ocean governance decisions and institutional mechanisms designed to make the best use of marine resources should take the relative values of different uses into account. When proposed uses of the resource conflict, such as the needs of recreational users and the dischargers of waste, the values at stake are not necessarily equal; closing an area to swimming because of pollution may have a much greater cost than closing an area to waste discharge because of swimmers. Similarly, restricting vessel traffic (or jet skiers) in the interests of wildlife viewers may have much lower costs than reducing whale or seabird populations in the interests of industry. Even though different users compete for precisely the same resource (marine waterways), the values they derive from the resource may differ greatly. Efficient and equitable governance depends on sound information about the economic values of the commercial and noncommercial uses of marine resources.

Moreover, governing institutions should reflect these economic values. Typically, commercial users of marine resources—whether for transportation, fishing, or mineral extraction—are well organized and well represented in governance decisions. Recreational users and other nonconsumptive users of marine resources are typically, though not always, less well organized and less well represented in governance institutions and are sometimes excluded from decision making processes. Such imbalances almost inevitably result in management decisions that do not lead to the best uses of marine resources. Governing institutions and processes should ensure that the most valuable uses of marine resources are effectively represented.

The values of marine goods and services, although they are accepted as real and important in meeting human needs, are systematically underestimated because they are public goods and are not priced in the market place. Despite continued warnings from scientists about the long-term consequences of the overuse and degradation of marine resources, society in general, and policy makers in particular, have been reluctant to address the problem of the overexploitation of marine resources (Costanza et al., 1997).

Unless some attempt is made to assign realistic values to marine resources and resource services, they will continue to be undervalued and used inefficiently. The rationale for estimating the monetary value (as opposed to the ecological value) of various marine resources is that they provide a means for deciding whether investments in the conservation, preservation, and management of these resources will improve the welfare of society. Knowing the value of resources and resource gives decision makers a better sense of the true costs of resource use and the true long-term benefits of preservation, conservation, restoration, or enhancement.

Unfortunately, there is a general misconception that economics may skew the debate over the conservation or restoration of marine resources in favor of developers and users; a corollary of this misconception is that economic analysis will only illustrate the extent to which laws requiring the protection of marine resources affect the economy through reductions in economic activity (e.g., tax revenues, employment and income). In reality, accurate information concerning the economic value (as opposed to the economic effects) of marine resources, functions, and services can support the argument in favor of preservation, conservation, restoration, and enhancement by providing a more complete and accurate picture of short- and long-term costs and benefits (Costanza et al., 1997).

Typically, economists have estimated the value of marine resources by concentrating on the components of the marine environment that have short-term obvious value to individuals (e.g., commercial and recreational fisheries, flood control by wetlands), which can be readily estimated with existing economic techniques. Frequently, values have been assigned to tangible, on-site resource services without regard to the underlying ecosystem. Bingham (1995) suggests that evaluating only those components of the ecosystem that are of immediate value to individuals, and focusing on short-term changes in the ecosystem, ignores changes in ecosystems that play out over time and space and that may be irreversible.

Garrett Hardin's essay on the tragedy of the commons is the best known description of the inefficiencies likely to arise from unrestricted access to a resource. He pointed out that open access undermines incentives for sustainable use of a resource (Hardin, 1968) because each user expects that other users will harvest as much as they can if he himself does not. Hardin's solution, that government must intervene on behalf of commonly-owned resources, disregards the ability of local communities to create effective and sustainable governance regimes for controlling resources.

Marine areas face an extreme version of the tragedy of the commons. The tradition of freedom of the seas is a tradition not only of free access—i.e., a tradition that the ocean is a commons for all mankind—but also of weak or unclear governing authority. In recent years, national governments have extended their jurisdictions 200 nautical miles offshore under the international concept of the EEZ, and most nations (but not the United States) have ratified and/or became parties to the 1982 United Nations Law of the Sea Convention. Still, formal governmental authority offshore tends to be weaker and less well defined than it is on land.

TRENDS IN MARINE AREAS

Several trends (Bacon, 1996) suggest that pressures on marine areas are intensifying and that severe consequences for people and the marine environment will result unless there are changes in the management regime. These trends include:

- the increasing power of humans to affect the marine environment through both mechanical means (fish harvesting, dredging) and chemical means (contamination by chemical substances, including petroleum and endocrine receptors)
- global climate changes and their effects on sea level, the intensity and frequency of storms, rainfall distribution, and recreation
- the growth of coastal populations

Although some studies have been done to estimate the value of recreational uses (including marine recreation) (Smith and Kaoru, 1990; Walsh et al., 1992), no database has been developed showing how the economic value of marine resources and resource services has changed over time. There are data, however, about contributions of coastal communities to the gross national product (GNP). One study, based on the assumption that the contribution of coastal communities to the GNP reflects a dependence on marine tourism, manufacturing, and transportation, came to the following conclusion:

> Based on payroll and employment, our estimates demonstrate that the coastal zone is a key economic sector that contributes more than 30 percent of the national GNP. Most of this value comes from the service sector, but even without that type of economic activity, the coastal zone accounted for some $55 billion in 1985. Our estimates also show that the coastal zone has become more important over time, growing from 30.1 percent of GNP in 1978 to 31.4 percent in 1985 (Center for Urban and Regional Studies, 1991).

An examination of individual components of marine economic activity bears out these findings. For example, fish landings have shown modest increases in value at certain times. However, a closer look at the available information suggests that this trend will not continue because many fish stocks have declined below the levels that support the maximum sustainable yield on a continuing basis (NOAA, 1997). The concern, therefore, is that the collapse of the New England groundfish industry, for example, could be a harbinger of the future of other fish stocks in the United States. One study has estimated, however, that if all fish stocks in the United States were managed properly, the potential net value of U.S. fisheries could be increased by $2.9 billion (Stroud, 1994).

Another indicator of change is that although the number of recreational fishing trips in U.S. waters remained fairly constant throughout the 1980s, the estimated catch declined substantially—by nearly 30 percent (Council on Environmental Quality, 1992). Another disturbing trend is the increase in the number of closings of shellfish beds—from 13 percent in 1966 to 31 percent in 1985 to 37 percent in 1990 (Weber, 1995).

Some trends in offshore oil leasing could also be disturbing in terms of the long-term, efficient utilization of marine resources. Since 1983, the percentage of offshore lands subject to moratoria on leasing has increased from about 3 percent to nearly 18 percent (Outer Continental Shelf Policy Committee, 1993). Although this trend reflects a genuine concern for protecting the marine environment, moratoria are a blunt governing instrument and may not result in the efficient long-term utilization of resources.

Trends in the discharge of wastes into the marine environment suggest that there has been some reduction in the discharge of certain pollutants in recent years. NOAA's Mussel Watch program, which test mollusks and other organisms for selected pollutants, found a decrease between 1986 and 1993 in several compounds, such as DDT, PCBs, and metals like cadmium. At the same time, the number and volume of oil spills and other toxic spills also appears to have decreased (Bacon, 1996).

An increase in discharges of nutrients from sewage treatment plants and agricultural runoff, and a resultant increase in the levels of eutrophication, is cause for concern. For example, Long Island Sound has shown steady deterioration for several decades, with areas of low dissolved oxygen increasing from 350 to 517 square miles just between 1985 and 1987. Similarly, the concentration of nitrate from agricultural runoff doubled in the Mississippi River from 1960 to 1980, resulting in widespread areas of low dissolved oxygen in the north central Gulf of Mexico (Weber, 1995). Related data seem to indicate an increase in the amount of human waste discharged into the marine environment, causing a steady increase in the number of beach closings by states (Bacon, 1996; NRDC, 1996).

The integrity and status of marine habitats are difficult to assess because of insufficient data; however, the available information suggests that the trends are negative in terms of the loss of wetlands (NRC, 1995c). For example, from the mid-1970s to the mid-1980s, there was a 71,000-acre loss in salt marshes and a 6,000-acre loss in sea grass beds throughout the coterminous United States (Bacon, 1996). A high proportion of the marsh loss is associated with the delta system of the Mississippi. Coastlines and beaches provide habitats for a wide range of marine life, including many threatened or endangered species. Natural erosion of these areas is aggravated by the construction of dams that impound sediment that would otherwise reach the shore and the blockage of naturally migrating inlets by structures that interfere with sediment transport (NRC, 1995a). Economic and environmental trends do not take place in isolation. There are strong synergistic effects among fisheries, habitat destruction, and pollution. Understanding the interactions among various human activities and developing comprehensive solutions is an important reason for improving marine area governance.

These interactions are well illustrated in the case of the Chesapeake Bay, where human induced distortions in the cycling of nutrients is a major problem. The primary sources of increased nutrients since 1950 have been human waste from urban areas and runoff from agricultural lands and atmospheric deposition to the bay and its watershed. Nutrients have caused excessive algae growth and subsequent die off, which by blocking sunlight and lowering the level of dissolved oxygen, respectively, have resulted in a loss of underwater vegetation (Boynton et al., 1995). This vegetation had been a crucial element in the overall regulation of the nutrient cycle on a seasonal basis. The loss of vegetation added to the imbalance, which was further distorted by the loss of oyster populations in the bay (for a variety of reasons including overfishing and disease) (NRC, 1994b). The loss of oysters, which are filter feeders, has further diminished the ability of the bay to process nutrients.

Ship discharging sewage sludge into the ocean. Photo courtesy of the Marine Board.

Shorebirds on marsh near Willapa Bay, Washington. Photo courtesy of Katharine Wellman.

The Chesapeake Bay is only one example of the complex interactions between humans and the natural environment. This example demonstrates that the incremental decline of several resources simultaneously can result in the virtual collapse of an entire system. It also suggests that efforts at restoration are often complicated and must necessarily be integrated across a range of issues (NRC, 1994c).

Conflicting Uses

Satisfying the diverse needs and wants of a large, rapidly growing coastal population is extremely difficult in many parts of the United States by the fixed supply of coastal resources and space, by the limited capacity of natural ecosystems to assimilate human stressors, and by legal systems of property rights that often do not differentiate between private and public ownership. The country is facing a new class of coastal and ocean management problems that can be characterized as conflicts over the use, or nonuse, of finite coastal environmental resources; development versus protection; and public interest versus private property rights. Sometimes, well functioning markets can handle these conflicts. Sometimes, however, markets do not accurately reflect the impacts of activities on the environment or on other users. This is especially true in the marine

Skipjacks (traditional oyster boats) on Chesapeake Bay. Photo courtesy of William Eichbaum.

environment, where many factors are not measured in dollars. For example, although the pollution of estuaries with sediment, point and nonpoint pathogens, toxic substances, and other pollutants all affect the growth and availability of fish and shellfish stocks, thereby reducing the productivity of commercial and recreational fishermen, the effects are not reflected in the marketplace.

In the past, most so-called "externalities" have been handled in the United States by sectoral regulatory regimes designed primarily to enforce health, environmental, and safety standards. These regimes, however, may not be adequate for sustaining marine resources, particularly in the face of uncertainties about the long-term values of resources and the impacts of various activities. Regulations are inherently reactive rather than proactive, and they may not mesh with the market signals that firms and individuals use to make decisions. They may also be ineffective for translating long-term goals into short-term incentives.

CURRENT MANAGEMENT REGIMES

A broad spectrum of coastal and marine issues must be considered for managing resources, safeguarding the health of the ecosystem and maintaining biodiversity, and providing a framework for using resources and space with a minimum of conflict. Managed areas range from small closed areas, or harvest refugia (designated to protect specific resources or habitats or to prohibit specific activities), to extensive coastal and/or marine areas for the coordinated management of many species, habitats, and uses.

A fully developed system that meets all of the objectives and contains all of the elements described in Chapter 1 necessarily evolve over time in response to actual experience. Nevertheless, based on experience with various marine management areas, steps to improve management can be taken now. This section sets forth approaches that are being tried now that could be starting points for developing a model for improved marine area management.

Coastal Zone Management Program

The Coastal Zone Management (CZM) program was established in 1972 to develop the capacity of states to plan for and manage coastal land and water resources. The program provides funding through NOAA to coastal states (including the Great Lakes states) and territories for the development and implementation of measures to conserve and develop coastal resources.

Major features of the legislation and program include the designation of the state as the lead party in designing and implementing CZM programs. The federal government provides two kinds of incentives to the states: (1) grants for the development and implementation of CZM programs; and (2) consistency—the promise that federal activities will be consistent with state CZM policies. The program is centered around process-related standards contained in the legislation (i.e., state CZM programs must regulate the uses of land and water that affect the coastal zone). Thirty states and territories (out of the 35 that are eligible) have established CZM programs covering about 94 percent of the U.S. shoreline; another four or five programs are in the development phase.

State CZM programs are required to include the territorial seas (out to the three-mile state offshore boundary). Hence, coastal states are already charged with planning and governing this important marine area. Recently, several states (notably Oregon, Hawaii, and California) have extended their CZM-related planning and policy-making area beyond the three-mile boundary into the waters of the 200-nautical mile EEZ. These initiatives into the offshore ocean do not yet have a formal legal basis, but they demonstrate strong interests on the part of most coastal states.

The existing CZM program provides a landward and coastal zone "anchor" for adjacent marine area planning and governance. A wealth of practical experience in marine area governance has been obtained by states that have participated in (and influenced) the federal offshore oil/gas program, the work of the regional fishery councils, and planning for marine sanctuaries. The CZM program has demonstrated the value of legal devices (like the federal consistency provision of the CZMA) for increasing the level of coordination and cooperation between states and the federal government in coastal and ocean management.

The CZMA of 1972 (as amended) includes a provision allowing coastal states to establish special area management programs. Relatively minor changes in the legislative language could enable coastal states to make initial efforts to implement the governance and management systems described in this paper. Other aspects of the CZMA offer vehicles for coordinating and integrating the decision making processes.

National Marine Sanctuaries

The National Marine Sanctuary (NMS) program was established in 1972 as part of the law regulating ocean dumping. The NMS program is administered under the National Ocean Service of NOAA. There are now 12 NMSs covering about 20,000 square miles of the U.S. ocean area and two proposed NMS sites (see Box 2-1).

The objectives of the NMS program are (1) to identify and designate areas of special national significance as sanctuaries, (2) to develop and implement coordinated protection and management plans for sanctuaries, (3) to facilitate public

BOX 2-1
Designated National Marine Sanctuaries and Proposed Sites

Site	Year Designated	Square Miles (Protected Area)
Key Largo, Florida*	1975	132
MONITOR (sunken ship), North Carolina	1975	1
Channel Islands, California	1980	1,658
Gray's Reef, Georgia	1981	23
Gulf of the Farallones, California	1981	1,255
Looe Key, Florida*	1981	7
Fagatele Bay, American Samoa	1986	0.37
Cordell Bank, California		526
Florida Keys, Florida	1990	3,707
Hawaiian Islands	1992	1,721
Humpback Whale, Hawaii		1,300
Stellwagen Bank, Massachusetts	1992	842
Flower Garden Banks, Texas	1992	56
Monterey Bay, California	1992	5,328
Olympic Coast, Washington	1993	3,310
Thunder Bay, Michigan	(proposed)	400
Northwest Straits, Washington	(proposed)	728

*Incorporated into Florida Keys in 1997

and private uses insofar as they are compatible with resource protection, and (4) to support scientific research and public education in the sanctuaries. The overriding policy is to protect "sanctuary resources," living or nonliving resources that contribute to the conservation, recreational, ecological, historical, research, or educational value of the sanctuary.

NOAA has considerable powers to implement this program. Activities prohibited within sanctuaries include any activity that would destroy or damage sanctuary resources, the sale or transport of sanctuary resources, violations of regulations or permits, and interference with enforcement. Special permits may be issued to allow otherwise prohibited activities, and NOAA may receive reasonable fees to cover costs. NOAA also has limited powers to review the activities of other federal agencies if they could endanger sanctuary resources.

Sanctuary officials have broad enforcement powers including the ability to impose civil penalties, forfeitures, and injunctions. The liability provisions of the program are especially important. Individuals or vessels that damage sanctuary resources must pay response and damage assessment costs and are liable for damages based on restoration costs or "value."

The process for designating ocean areas as sanctuaries has followed two routes. The first NMSs were designated after elaborate reviews of candidate sites based on specified factors and agency consultations. More recently, Congress itself has designated a number of sites after finding it to be in the national interest to have a NMS in a particular location. The Florida Keys NMS is a good example. Congress passed a special law establishing the Florida Keys NMS and added special features, such as precise marine boundaries, a water quality planning process, and rules for controlling vessel traffic. Other congressionally designated sanctuaries include the Monterey Bay, the Olympic Coast, Hawaiian Humpback Whale, and Stellwagen Bank sanctuaries.

Box 2-1 shows the 12 existing national marine sanctuaries and the three active candidates, indicating the year of designation and size in square miles. The 12 designated sanctuaries can be divided into five types: historic preservation (Monitor); reefs (Looe Key, Key Largo, Fagatele, Gray's Reef, Flower Garden); banks (Cordell, Stellwagen); islands (Farrallones, Channel, Hawaiian); and ocean areas (Florida Keys, Monterey Bay, Olympic Coast).

Regulations for each sanctuary specifying prohibited activities vary but tend to fall into common categories. Exploring for and developing oil and mineral resources tend to be excluded. Discharges or deposits of material are disallowed unless they are part of traditional fishing operations or are approved under specific regulations. In some sanctuaries, cargo carrying vessels may not navigate within prescribed distances from islands. Drilling into, dredging, or altering the seafloor is usually prohibited, with some exceptions for anchoring, navigational aids, certain fishing operations, and others. Aircraft must stay above 1,000 feet in certain zones over or adjacent to a sanctuary to protect marine mammals and seabirds. Historical resources may not be moved or removed, and certain wildlife

View of Florida Keys from the air. Photo courtesy of William Eichbaum.

may not be taken or possessed, especially species protected under other laws. The Florida Keys NMS has proposed regulations that go even farther, including identifying zones where certain uses are allowed or prohibited (for example fishery replenishment zones, restoration zones, research-only zones, and facilitated-use zones).

Three issues have been prevalent in the development of the NMS program. The first relates to vessel transportation. Sanctuary managers have attempted to reduce the risk of spills and accidents from tankers and other larger cargo vessels by proposing the establishment of "Areas to be Avoided," vessel exclusion zones, and vessel traffic schemes. The maritime industry (and to some extent the U.S. Navy) has resisted these restrictions on the basis of added costs and unproven risk reduction. A related transportation issue is the controversy over jet ski operations in an NMS. A federal court decision upheld the power of NOAA to limit jet skis to prescribed zones within the Monterey Bay NMS. This issue has reached national proportions, and lawsuits are under way between the personal watercraft industry and local, state, and federal government agencies.

A second issue relates to the control of fishing within sanctuaries. The NMS statute limits the power of NOAA to impose fishing regulations unless the regional fishery management councils defer to the sanctuary. Some sanctuaries have attempted to protect fish habitats, such as shallow banks and reefs. Conservationists argue that the NMS should be authorized to deal with what they consider to be one of the biggest ocean resource issues—the decline of fisheries stocks. In the Florida Keys NMS, fisheries replenishment zones have been proposed where fishing would be restricted to protect fish populations in the area. This proposal is controversial and may or may not be implemented.

Third, the NMS program faces the challenge of integrating the demands of many governmental and private interests, a formidable job for small NMS staffs. Nevertheless, progress is being made. The statute calls for state governments to review federal regulations and management programs when state-owned waters are involved (which is almost always the case). In some cases (e.g., Florida Keys NMS), federal and state officials "co-manage" sanctuary resources. The emergence of sanctuary advisory committees at some NMSs has brought private and local government interests into the management process.

Perhaps the most difficult task of integration is among federal agencies. For example, a sanctuary manager may need to coordinate with adjacent national park managers regarding visitation rules with the Federal Aviation Administration regarding flight restrictions, with the U.S. Navy regarding restrictions on naval operations, with the U.S. Coast Guard regarding reporting vessel activity or other navigational issues, and with the National Marine Fisheries Service (NMFS) regarding fishing restrictions and marine mammal protection.

The NMS program offers many opportunities for improving governance. Although the primary objective of the sanctuaries program is to protect exceptionally valuable marine resources, the process could be used as a model for improving management and governance beyond the sanctuary boundaries.

National Estuary Program

An estuary is a coastal area where fresh water from rivers and streams mixes with salt water from the ocean. Many bays, sounds, and lagoons are estuaries. Estuaries provide safe spawning grounds and nurseries and are critical for fish, birds, and other wildlife.

Congress established the NEP (National Estuary Program) as part of the Clean Water Act to protect and restore estuaries while supporting economic and recreational opportunities. The EPA designates local NEPs to develop partnerships among the government agencies that oversee estuarine resources and the people who depend on those resources for their livelihood and quality of life. Through a consensus-based process, stakeholders work together to develop a plan of action that meets the needs of their own communities. To date, 28 local NEPs have found practical and innovative ways to revitalize and protect their estuaries.

The 28 estuary programs listed in Box 2-2 are in various stages of development. Some are developing management plans; others are already implementing management plans. Currently, 10 programs are implementing management plans, ranging from habitat restoration to septic tank conversion. The key element of all local NEPs is public involvement (EPA 1995). Under the NEP, the administrator

BOX 2-2
National Estuary Programs

PROGRAMS

Albemarle-Pamlico Sound, North Carolina	Massachusetts Bays, Massachusetts
Baratarla-Terrebonne Bays, Estuarine Complex Louisiana	Morro Bay, California
	Mobile Bay, Alabama
Barnegat Bay, New Jersey	Narragansett Bay, Rhode Island
Buzzards Bay, Massachusetts	New Hampshire Estuaries,
Casco Bay, Maine	New Hampshire
Charlotte Harbor, Florida	New York-New Jersey Harbor,
Columbia River, Washington and Oregon	New York and New Jersey
Lower Corpus Christi Bay, Texas	Peconic Bay, New York
Delaware Estuary, Delaware, Pennsylvania, and Peconic, New Jersey	Puget Sound, Washington
	San Francisco Bay, California
Delaware Inland Bays, Delaware	San Juan Bay, Puerto Rico
Galveston Bay, Texas	Santa Monica Bay, California
Indian River Lagoon, Florida	Sarasota Bay, Florida
Long Island Sound, New York and Connecticut	Tampa Bay, Florida
	Tillamook Bay, Oregon
Maryland Coastal Bays, Maryland	

of the EPA is authorized to convene management conferences for the following reasons:

- to assess trends in water quality, natural resources, and uses of an estuary
- to identify the causes of environmental problems in an estuary
- to relate pollutant loads to observed effects on the uses, water quality, and natural resources of an estuary
- to develop a comprehensive conservation and management plan (CCMP) that recommends corrective actions and implementation schedules
- to develop a plan for coordinating the implementation of the CCMP among federal, state, and local agencies
- to monitor the effectiveness of actions implemented under the CCMP
- to ensure that federal assistance and development programs are consistent with the CCMP

The NEP operates on the principle that all components of the environment are interconnected. The traditional focus on problems with specific resources has not solved resource and water quality problems. Estuarine management requires dealing with a variety of laws, management initiatives, and funding from numerous public and private sources.

To reconcile the many interests and resource problems in estuarine areas, the NEP has adopted a consensus-building process that attempts to coordinate activities among a broad spectrum of stakeholders. The process seeks to build partnerships among all levels of government, the private sector, and the public. The NEP provides a forum where stakeholders (government agencies, industry, environmental groups, and the public) can work together to develop strategies for resource management. EPA participates as one of the stakeholders.

The NEP was established because conventional pollution control programs were not adequate for dealing with complex estuary problems. The NEP approach emphasizes partnerships among all interested parties (national, state, and local governments, as well as nongovernmental interests) and consensus-based decision making based on scientific information. But management conferences do not have regulatory authority so recommendations must be implemented by existing authorities at the federal, state, and local levels.

The NEP is an attempt to integrate all aspects of ecosystem management of near-shore waters. The program is area-based, addresses problems that affect entire ecosystems, including watershed and marine issues, and is composed of government agencies from federal, state, and local levels and other relevant constituencies. The biggest drawback of the program is that it does not provide funds or processes for implementation or accountability.

Essex Estuary, Buzzard's Bay, Massachusetts. Photo courtesy of William Eichbaum.

Outer Continental Shelf Oil and Gas Leasing Program

The federal OCS (outer continental shelf) oil and gas leasing program has been a source of substantial revenue to the nation while providing a significant portion of the domestic oil and gas supply. Since 1954, the federal government has received more than $100 billion from the leasing and production of OCS oil and gas. Production from the OCS has consistently provided about 12 percent of domestic oil production and more than 20 percent of domestic gas production. The MMS (Minerals Management Service) estimates that about one-third of future domestic oil discoveries and almost one-half of future domestic gas discoveries will come from the OCS. In addition, very large undiscovered fields (greater than 100 million barrels) probably lie in the deep water (water depths greater than 1500 feet) in the Gulf of Mexico OCS (OCS Policy Committee, 1993).

In spite of the benefits and importance of the OCS to the domestic energy supply, the OCS program has long been a source of controversy. The conflicts have increased since 1982 and involve federal agencies, Congress, various states, local governments, environmental groups, the energy industry, and private citizens. Overall, the controversies and the lack of measures for dealing with them have seriously diminished the effectiveness of the program.

The Outer Continental Shelf Lands Act (OCSLA), enacted in 1953, was the culmination of years of dispute between the federal government and the states over primary responsibility for coastal and offshore waters. The act authorized

Offshore oil production platform, Gulf of Mexico. Photo courtesy of the Marine Board.

the Secretary of the Interior to lease federal offshore lands for mineral exploration, development, and production. It also gave the secretary a mandate to develop OCS oil and gas resources, but it provided for very limited input or influence from the states. Governance under the act was reasonably effective during the 1950s and 1960s when leasing and production were focused off the Texas and Louisiana coasts where the oil and gas industries had long been important to the local economy. Expansion into offshore waters seemed to be a natural extension of existing activities and was supported by an infrastructure that was already in place.

Leasing federal lands along the Pacific Coast began in the early 1960s. After several years of exploration, a major oil reserve was discovered in the Santa Barbara Channel. In 1969, there was a blowout and oil spill at a platform in the channel. The ensuing environmental damage and the perceived arrogance of the responsible oil company attracted national attention. As a result, opposition to offshore oil and gas leasing spread.

After the Santa Barbara incident, support for more stringent environmental safeguards and for giving states a stronger role in activities off their coastlines grew rapidly. The National Environmental Policy Act (NEPA) was passed in 1969, and the CZMA was passed in 1972 and amended in 1976. These two acts gave the states their first real voice in OCS decisions, although the primacy of the federal government was preserved.

In response to the Middle East oil embargoes in 1973 and 1974, plans for a significant expansion in OCS leasing were developed. Numerous lease sales were planned in frontier areas, including Alaska, the Atlantic Coast, and the Pacific Coast, as well as in the producing areas in the Gulf of Mexico. This accelerated program increased environmental concerns. In 1978, the OCSLA was amended to give more consideration to environmental factors and to allow more state and local involvement in OCS decision making. However, once again, the federal government retained the right to accept or reject recommendations from state and local governments.

In 1979, another disruption in oil supplies renewed the impetus to accelerate OCS development, which, in turn, raised further widespread concerns about potential negative environmental and socioeconomic impacts, especially in frontier areas. After the Secretary of the Interior greatly increased the area proposed for offshore leasing, states and other groups turned to the annual congressional appropriations process as a means of influencing and controlling OCS activities. Starting in 1982, Congress enacted a series of one-year moratoria prohibiting the Department of the Interior from engaging in any activities in certain geographic areas (mostly offshore California). Between 1982 and 1993, the acreage covered by these moratoria grew from 0.7 million acres to 266 million acres. As a result of these moratoria, as well as economic factors, most offshore leasing has been concentrated in the Gulf of Mexico (although some areas offshore California and Alaska are being developed under leases acquired prior to the moratoria).

There are several opportunities for public participation throughout the federal leasing process. The current system provides opportunities for making decisions based on consultation and cooperation, but in practice this process has sometimes led to protracted controversies and conflicts especially along the West Coast.

Fisheries Management

Living marine resources currently support extensive commercial, recreational, and economic activities in all coastal areas. In 1995, commercial landings by U.S. fishermen were 9.9 billion pounds, valued at a record $3.8 billion. The 1995 U.S. marine recreational finfish catch was estimated at 339.1 million pounds taken on 65.5 million fishing trips (NOAA, 1996). These are just a few of the benefits Americans derive from living marine resources.

However, many marine species are under stress from overexploitation or habitat degradation or both. More than a third of all fish stocks for which we have reliable information are overutilized, and nearly half are below optimal population levels (NOAA, 1997). Some marine mammals, turtles, and fish are in danger of extinction, and many more are threatened. Maintaining and improving the health and productivity of these species is essential to the sustainable use of marine resources as well as to the health and biodiversity of marine ecosystems. Many factors, both natural and human-related, affect the status of fishery stocks, protected species, and ecosystems.

The Secretary of Commerce, through NOAA's NMFS (National Marine Fisheries Service), is responsible for the conservation and management of most of the living marine resources within the 200-nautical mile limits of the U.S. EEZ. NMFS also plays a supporting and advisory role in the management of living marine resources in coastal areas under state authority. Management and conservation plans are developed through extensive discussions with state and tribal officials, other federal agencies, fishermen, processors, marketers, public interest groups, universities, and the general public, as well as through partnerships with international science and management organizations.

The 1976 Fishery Conservation and Management Act (FCMA), under which fisheries within the EEZ are regulated, established eight regional fishery management councils to prepare fishery management plans for the nation's fisheries. Members of these councils are appointed by the Secretary of Commerce, based on recommendations from governors of the states in each region, and represent diverse interests. Each council also includes nonvoting members from the U.S. Fish and Wildlife Service, the U.S. Coast Guard, and the U.S. Department of State. Fishery management plans developed by the councils are subject to approval by the Secretary of Commerce. In some cases, fishery management plans are developed directly by NMFS, with advice and comment from the public, including the regional councils.

Shrimp trawler, Appalachacola Bay, Florida. Photo courtesy of William Eichbaum.

NMFS is responsible for preserving protected marine species through the Endangered Species Act, which protects species from extinction, and the Marine Mammal Protection Act, which promotes the maintenance of marine mammal populations at optimal levels. Under various statutes, including the Fish and Wildlife Coordination Act, NMFS can mitigate losses or damage to habitats vital to living marine resources. For example, NMFS reviews federal proposals that may affect habitat vital to living marine resources and makes recommendations for the adequate conservation of those resources. Recent revisions to the FCMA mandate an essential fish habitat program, and the Endangered Species Act and Marine Mammal Protection Act mandate the conservation of critical habitats of threatened and endangered species and marine mammals.

The FCMA was the first major act to establish a decentralized marine management structure, and the system of regional councils remains firmly entrenched. The implementation of fishery management plans reinforces the decentralized regional council system.

Many of the problems that surfaced early in the implementation of the FCMA persist in varying degrees: continued overfishing of certain stocks; an adversarial relationship between some of the fishery councils and the NMFS; conflicts among user groups; the vulnerability of the fishery management process to delays and political influence; a lack of accountability; inconsistencies in state and federal management measures; and the adoption of unenforceable management measures (NRC, 1994a).

A recent report by the Ocean Studies Board of the NRC (NRC, 1994a) identified four topics that Congress should address when restructuring the FCMA: the overfishing (including the amount of commercial fishing and the definition of optimum yield); institutional structure (including management of fishery councils); the quality of fishery science and data; and an ecosystem approach to fishery management. To clarify lines of authority, the report found that the regional fishery councils should continue to bear responsibility for allocating and capitalizing controls for fisheries within their regions. The establishment of acceptable biological catches, however, was considered as a scientific determination that could be best made by having NMFS, state agencies, and other interested scientists provide their estimates of appropriate catch levels to the councils. A scientific advisory committee for each fishery council could then determine the acceptable catch levels so that the maximum yield of stocks could be sustained over the long term.

The 1996 reauthorization and amendments to the FCMA addressed the four recommended topics. The basic institutional structure and role of the regional fishery councils in the preparation of fishery management plans and proposed regulations was retained. The qualification for membership on the regional councils was not changed; however, the conflict of interest standards were tightened. The definition of optimum yield was changed to require rebuilding overfished stocks. The regional councils are now responsible for preparing plans, amendments, or regulations to end overfishing and to replenish stocks within 10 years. If a council fails to meet the one-year deadline, the Secretary of Commerce has

Local salmon trawlers, Deception Pass, Whidbey Island, Washington. Photo courtesy of Katharine Wellman.

the authority to adopt interim measures to address overfishing in any fishery. The amendments do not make clear, however, if the Secretary has the authority to act without a majority vote of a council.

The amendments present a mixed picture with respect to controls on capitalization. They specifically authorize using individual fishing quotas (IFQs) to control fishing and catches. However, they impose a moratorium until October 1, 2000, on the Secretary's approval or implementation of new programs for assigning individual fishing rights. The amendments require the National Academy of Sciences to prepare a comprehensive report on IFQs by October 1, 1998, including recommendations for a national policy. The new legislation also establishes a program for identifying and protecting essential fish habitats and requires the regional councils and NMFS to assume a more active role in this area. The amendments also require steps to minimize fish mortality from by-catch.

In summary, the 1996 amendments clearly recognize the need for better governance and management, and they provide a basis for improving governing institutions and management tools. Improvement, however, will depend on the willingness of fishery councils, the NMFS, the states, and the fishing industry to break old patterns of operation. The recommendations for improved governance and management in Chapters 5 and 6 of this report suggest how this might be accomplished.

State Ocean and Coastal Management Programs

Changes in U.S. laws and policies in the last decade have given states greater influence over ocean use and activities. At the same time, states have taken steps to increase their capacity for ocean management through policy development, and in a few cases, through policy implementation. The trend toward increasing state involvement will probably continue because many new vehicles for state participation in ocean affairs are now in place.

Growing state influence over ocean use can be observed in five categories: coastal zone management, oil pollution control, OCS oil and gas development, governance of NMSs, and fisheries management. In each category, the role of the federal government is still important, sometimes even primary, but the influence of state and local interests and their areas of jurisdiction have increased.

Under the CZMA, state governments are encouraged to develop programs for controlling land and water uses in coastal areas and three miles seaward. One of the powers given to the states, called the "federal consistency" power, gives state governments more authority to review activities at the federal level that affect land, water, or natural resources, even if they occur well outside of the state's coastal zone. The President can overrule a state, but only if a judicial ruling has been made that consistency is not possible. This means that states now have considerable influence over federal activities in the ocean well beyond the three-mile boundary (Eichenberg and Archer, 1987).

The second category of increasing state influence over ocean activities is oil pollution control. By 1978, there were four major laws establishing oil pollution control, cleanup, liability, and compensation.[2] But the key issue was whether the new rules should preempt states' rules. The Exxon Valdez oil spill in 1989 led to the passage of the Oil Pollution Act of 1990 (OPA 90) which preserved state oil pollution laws (Mitchell, 1991). Thus, the federal government is responsible for establishing minimum standards for oil transportation and safety, but many states have established their own standards that exceed those of the federal government. Many states impose penalties and fines and have established requirements for liability insurance, the removal of obstructions and pollutants, and the establishment of trust funds.[3]

The third category that illustrates growing state influence over ocean use is in the area of OCS exploration, leasing, and development. In the last decade, the United States has experienced what might be called "a 10-year war" between the lower 48 states and the federal government over OCS oil and gas development (Hershman et al., 1988). Many of the proposed leases were highly controversial and were adamantly opposed by the states. Ultimately, state interests won out over federal interests because the authority of the MMS was undermined by a variety of factors, such as congressional moratoria, special requirements added to appropriation bills, special study requirements, requirements for particular procedures for environmental impact statements, and others (Fitzgerald, 1987; Kitsos, 1994; Lester, 1994).

The fourth category relates to the federal NMS program. Although national NMSs are designated, staffed, and partly funded by the federal government, many of them have been established in response to strong local pressures, coupled with congressional influence over the designation process. NMSs have been used as vehicles for resolving local environmental problems, such as conflicts over oil and gas development, shipping, and the disposal of dredged material. In other words, the objectives of establishing the sanctuaries are often local or regional, rather than national.[4]

----

[2]The Federal Water Pollution Control (Clean Water) Act Amendments of 1972 (33 U.S.C.S. §§ 1251 et seq., see § 1321); The Deepwater Port Act of 1974 (33 U.S.C.S. §§ 1501 et seq., see § 1509, 1517 (liability)); Outer Continental Shelf Lands Act Amendments of 1978 (43 U.S.C.S. §§ 1331 et seq., see § 1341 (a)-(d)); Trans-Alaska Pipeline Authorization Act of 1973 (43 U.S.C.S. §§ 1651 et seq., see § 1653).

[3]In 1990, the state of California enacted the Lempert-Keene-Seastrand Oil Spill Prevention and Response Act, which set up state vessel inspection programs, oil spill contingency plans, an Oil Spill Response Trust Fund, and an Oil Spill Prevention and Administrative Fund to assist in oil cleanup in state waters (Cal. Gov. Code @ 8670.1 et seq.). The Washington state legislature established a marine safety office in 1991, an agency of the state government, responsible for promoting the safety of marine transportation in Washington (RCW 43.21I.010)

[4]The Flower Garden Banks National Marine Sanctuary, located 120 nautical miles south of Cameron, Texas, was designated in November 1991. It specifically prohibits oil and gas development (P.L. 102–251, Title 1 § 101, 106 Stat. 60). The Stellwagen Bank National Marine Sanctuary was

The fifth and final category illustrating the increase in state influence over ocean use is fisheries management. The FCMA was the first major marine policy to establish a decentralized management structure that included state officials, as well as experts nominated by governors of the states in the councils. The system of regional councils has been firmly entrenched since 1977, and the implementation of fishery management plans has reinforced the decentralization of the regional council system.

Changes in the States' Capacity to Manage Marine Areas

The growing influence of state governments over ocean use is one measure of change in U.S. ocean governance and management policy. Another measure is the "capacity" to manage. Capacity is defined as the ability and commitment of a jurisdiction to develop, staff, and sustain institutions that deal with marine policy issues (King and Olson, 1988). In the past 10 years, the capacity of some state governments has grown, and in one case at least, the capacity of local governments has grown as well (Hershman et al., 1988).[5] Federal management capability is still substantial but, in some sectors, is not growing and could be declining relative to the recent increase in state and local responsibilities imposed by new laws.[6]

Ten states[7] might be considered "activist" states because they have initiated measures in the past decade to improve their capacities for ocean management and have attempted to define a greater role for themselves in national decisions concerning ocean resources. Some activist states have expanded established programs, such as the CZM program, and some have undertaken new initiatives.

---

designated by Congress in November 1992, encompassing 638 square nautical miles of federal waters situated on and around the bank, 6.3 nautical miles north of Cape Cod, Massachusetts. It specifically prohibits sand and gravel mining (P.L. 102–587, Title II, Subtitle B, § 2202, 106 Stat. 5048). Monterey Bay National Marine Sanctuary was designated by Congress in September 1992, encompassing an area approximately 4,024 square nautical miles off the coast of central California. It specifically prohibits oil and gas development (P.L. 102–368, Title I, § 102, 106 Stat. 1119). The Olympic Coast Marine Sanctuary was designated in August 1994; oil and gas development are specifically prohibited (P.L. 100–627).

[5]Santa Barbara County, California, has been active in overseeing many aspects of offshore oil development, and its staff and budget have increased.

[6]Since 1980, federal funding for marine-related programs has steadily declined. In particular, funding has decreased for agencies like the National Oceanic and Atmospheric Administration, which controls coastal zone management, marine sanctuaries, and fisheries management programs and plays in important role in the oil pollution and outer continental shelf oil and gas programs. Recently, proposals to eliminate the U.S. Department of Commerce have been introduced in Congress, including eliminating NOAA and redistributing many component programs (H.R. 1756, 104th Congress, 1st Session, 1995).

[7]Alaska, California, Florida, North Carolina, Hawaii, Maine, Massachusetts, Mississippi, Oregon, Washington.

Several trends have contributed to the increasing importance of state ocean policies. The first and most important was the aggressive state response to OCS leasing policies, which were a major source of contention between federal and state governments. States used their powers under OCSLA and CZMA, as well as in the congressional budget process, to force reconsideration of federal administrative actions. Under current arrangements, states shoulder many of the environmental risks associated with OCS development but share few of the financial benefits. The level of controversy was greatly reduced by a policy shift in June 1990, which scaled back the OCS program and delayed the sale of many leases.

The second trend involves the development of state-level policy reports and studies defining ocean management issues. Because states felt that the federal government was only concerned with oil, gas, and minerals development, they attempted to step in and fill the vacuum. In the mid-1980s, North Carolina published a policy report (North Carolina Marine Science Council, 1984) and held a follow-up workshop. Oregon (Good and Hildreth, 1987) and Washington (Butts, 1988) soon followed North Carolina's lead. In the late 1980s, Hawaii (Hawaii Ocean and Marine Resources Council, 1988), Florida (Christie, 1989), and California[8] issued a second wave of studies and plans. Alaska was added to the list after the completion of a comparative study of West Coast states (Cicin-Sain, 1990). Mississippi (McLaughlin and Howorth, 1991) and Maine (Catena, 1992) are the latest states to release ocean policy reports. All 10 policy reports explore models for ocean management that would increase their capacity and prepare them for a policy dialogue with the federal government.

Finally, state ocean policy initiatives are part of a larger national trend of states taking the lead in many policy initiatives (Osborne, 1988; Bowman and Kearney, 1986; John, 1994). Throughout the 1970s, states were asked to take on greater responsibilities for implementing federal legislation, such as the Clean Air Act of 1970, the Clean Water Act of 1972, and the CZMA of 1972. In the 1980s, changes in administrative policy shifted responsibility from the federal government to state governments in response to pressures to reduce the federal budget. This trend has resulted in increased state institutional capacity to address problems and an overall decentralization of policy making in the United States (Bowman and Kearney, 1986).

Although the state initiatives indicate interest and enthusiasm for ocean planning and management on the part of many states, the results have been modest. Efforts by the states have been intermittent and inconsistent. Many have been characterized by organizational changes and funding problems. For example, North Carolina, Florida, California, and Hawaii have shifted responsibility for ocean planning among various state agencies; Maine and North Carolina have simply updated existing policy reports after many intervening years; Missis-

---

[8]California Ocean Resources Management Act of 1990. Cal. Pub. Res. Code 36,000 et seq.

sippi, Florida, and Washington have done virtually nothing since 1991. Delays and redirected efforts can be attributed partly to changes in the OCS oil and gas leasing policy for the lower 48 states. In 1990 and 1991, the federal government delayed the sale of OCS oil and gas leases for a decade, removing the sense of urgency felt by Washington, Florida, Massachusetts, Maine, and North Carolina (each of whom faced the prospect of offshore development) and slowing policy planning.

States have also depended heavily on existing federally-funded programs to develop their ocean-related policies. Each of the states discussed above has a federally approved CZM program, which provides some funding for ocean planning, and more importantly, gives the state federal consistency powers. In addition, six of the ten states have offshore NMSs and participate extensively in the NMS program (which also provides substantial funding for sanctuary management, research, and education). Thus, although states are trying to shift policy away from the federal level, the vehicle for change is an existing federal program. This has important implications for the future.

Because policies are still being developed, it is difficult to see trends clearly. But it is clear that environmental protection is being given priority over oil and gas development, marine mining, ocean waste disposal, vessel movements, aquaculture, and other intensive or industrial uses. Oregon and Washington have adopted broad policies favoring renewable over nonrenewable resources. Washington, California, and Florida have convinced federal agencies to ban oil and gas development in NMSs. Massachusetts and California have placed tight restrictions on the disposal of dredged materials and other wastes in the marine sanctuaries off their shores. Furthermore, California and Florida have adopted zoning controls to limit jet skiing.

Existing fisheries have been given protected status in state-level ocean management policies. For example, even sophisticated ocean management plans like the Oregon Ocean Plan do not propose changing fisheries policy or management. Even though substantial issues have been raised concerning allocation, conservation, by-catch, and impacts on the ecosystem, states have generally been unwilling to address these issues outside the established fisheries management institutions. However, there may be small signals of change. In 1990, for example, California passed a fish-sanctuary law that prohibits fishing within four small harbor areas (Hildreth, 1995). States have also responded to the concerns of recreational fishermen, who often press for restrictions on commercial fisheries.

Even though policy development among the states has been slow and distinct trends are hard to discern, the states now play a significant role in ocean policy making. State roles have been institutionalized in at least three ways—through state CZM programs, participation in the development and staffing of NMSs, and the emergence of regional groups, such as fisheries management councils and oil pollution task forces. This institutionalization suggests that state policies will continue to influence how the ocean is used in the future.

NEED FOR BETTER GOVERNANCE

Despite the many programs and regulations that affect coastal and marine resources, areas, and activities, there are no basic principles or processes for establishing authority and accountability in the management of marine resources and the uses of ocean space. In other words, there is no coherent national *system*. The United States tends to manage its ocean resources and space on a sector-by-sector regulatory basis. One law, one agency, and one set of regulations may be applicable to a single-purpose regime (e.g., oil and gas development, fisheries, water quality, navigation, or protecting endangered species), and a single *ocean area* may be subject to a plethora of regulatory management regimes.

The single-purpose, overlapping, uncoordinated laws that generally characterize the present system in which various local, state, and federal agencies manage ocean resources do not account for the effects of any single activity on other resources or the environment, assess cumulative impacts, or provide a basis for resolving conflicts. In the absence of an overarching governance system, parties seeking to use ocean resources and space for economic purposes and parties concerned with environmental preservation often reach a stalemate. The delays inherent in this approach often have significant societal and economic costs.

The fragmentation of governmental agencies and responsibilities is both horizontal and vertical. At the present time, management of the marine environment is carried out at local, state, regional, and national (and, in some cases, international) levels of government. At any given level, various functions are carried out by a wide array of separate agencies and organizations, with limited or sporadic coordination. As a result, many situations are poorly or inefficiently managed, and conflicts can be solved with great difficulty, if at all.

Sometimes this fragmentation means that important issues, rather than receiving too much attention, fall through the cracks of various jurisdictions. For example, although a number of agencies purport to exercise partial responsibility for the management of marine habitats, the question of habitat protection as a whole may simply not be addressed. Fragmentation also means that real or potential conflicts either among governmental requirements or among proposed users are often not anticipated, and when they emerge, they cannot be resolved effectively. In the absence of a coherent, coordinated system, opportunities are lost and resources are squandered.

The environmental and economic health of the nation's marine areas are linked at the individual, community, state, national, and international levels. The interdependence of the economy and the environment are widely recognized by government agencies, economic users of the ocean, and the general public. The nation has moved beyond the early concepts of health, safety, and pollution control as added costs of doing business to a concept of broader stewardship, recognition that economic and social prosperity would be meaningless if the coastal environment is compromised or destroyed in the process (President's Council on Sustainable Development, 1996).

Despite the history of sector-based management of marine resources, some legislation, including the CZMA, the CWA, the OPA 90, admonish federal agencies, as well as states, to preserve, protect, develop, enhance, and, where possible, restore the resources of the U.S. coastal zone. In the din of conflicting voices advocating different interests, one can sometimes hear a faint call for balance. Better marine governance would include the recognition of conflicts among users in marine areas. Better understanding and recognition of property rights, the value of public goods, and the links between the ecological and economic systems could lead to more efficient use of resources and the expansion of nonintensive uses, such as ecotourism and recreation.

3

Lessons Learned

The preceding chapters defined the fundamental interests and values at stake in the governance and management of marine areas, which are constantly affected by threats, trends, and opportunities that undermine the effectiveness of traditional, fragmented, sector-by-sector governance and management systems. As a basis for recommending improvements, the committee examined a wide range of real-world situations, three of them in-depth (as case studies) and seven in less detail. This chapter summarizes lessons learned from these investigations, especially in terms of the organizational and behavioral aspects of governance and management systems and the tools necessary for effective management.

CASE STUDIES

The committee conducted in-depth examinations of three representative examples of marine governance and management processes and structures in situations where problems have been especially difficult to resolve: (1) the designation of marine sanctuaries and parks, (2) the resolution of multiple-use conflicts, and (3) fisheries management. Case studies were conducted by the Center for the Economy and the Environment of the National Academy of Public Administration (NAPA) and the Marine Policy Center of the Woods Hole Oceanographic Institution (WHOI) with oversight by the committee. (The complete case studies are available from the Marine Board.) The case studies focused on the following geographic areas:

- Florida Keys National Marine Sanctuary (NAPA)
- Southern California coast (offshore and coastal region from the San Luis

Obispo/Monterey County Line in the north to the U.S./Mexico border in the south) (NAPA)
• Gulf of Maine/Massachusetts Bay (WHOI)

In each of the case studies, an activity or issue of local importance was examined to reveal the success or failure of existing management and governance structures and processes. Although other concerns might also have been of interest to this study, the committee limited its focus to the core problems, use conflicts, fragmented decision making and management, and lost opportunities and economic benefits: offshore oil and gas leasing and production in the southern California case, fisheries management issues in the Gulf of Maine study, and problems of regional marine and coastal planning in the Florida Keys case.

The committee attempted to determine whether present systems succeeded or failed in managing conflicts, exercising stewardship by protecting the resource base and the broader environment, and realizing the potential economic benefits of appropriate resource development. The following questions were central to each case study:

• Were issues of long-term national interest identified and addressed?
• What was the role of intergovernmental (i.e., local, state, and federal) relations?
• To what extent were these characterized by coherence or fragmentation, cooperation or conflict?
• What was the nature of the institutions involved?
• What kinds of behavior (cooperative or adversarial, hierarchical or collaborative) characterized them?
• What kinds of conflict management were used (litigation, stalemate, compromise, partnership)?
• Were institutions flexible enough to evolve to meet changing needs or changing perceptions?
• Were environmental and economic goals appropriate?
• Were these goals achieved?
• What was the range of factors that entered into decision making?
• What was the range of stakeholder involvement in decision making?

Criteria for Conducting and Analyzing the Case Studies

In addition to the relatively open-ended questions listed above, the committee outlined the characteristics of successful governance systems. These characteristics are drawn from the background paper for the study (Appendix B), issue papers prepared by several committee members at the start of the project, and experience with coastal governance and management. The principles are defined in Chapter 1.

The following sections summarize the examples the committee examined, identify themes that appear in all or most of them, and compare these themes to the governance criteria listed above. To facilitate their evaluation, the committee summarized the examples in terms of:

- why each was considered a success (or failure)
- characteristics of the governing process
- enabling factors or prerequisites that set the stage for success (or failure)

Although the examples differ widely in terms of scale, scope, and time frame, they all reflect attempts to address the core governance problems of use-conflicts, fragmented decision making and management, and lost opportunities and economic benefits.

Florida Keys National Marine Sanctuary

The Florida Keys are a unique chain of islands extending south and west from the tip of the Florida peninsula. The reef (the only living coral reef in the continental United States) and the flats surrounding the Keys are home to a large number of marine species. The Keys are heavily used for tourism and commercial and recreational fishing, are adjacent to marine transport routes between the U.S. mainland and Latin America and the Caribbean, and are close to areas being considered for oil and gas exploration. The natural systems of the Keys may be affected not only by activities in the Keys themselves (e.g., diving, treasure salvage, development), but also by the ecological problems of the adjacent Florida Bay. In 1990, in recognition of the environmental value of the Keys and apparent threats to their health, Congress declared the Keys a NMS and directed NOAA to complete a sanctuary management plan to protect them.

This plan is far more extensive than plans for other marine sanctuaries. It covers the entire Florida Keys and adjacent waters and details responsibilities for 18 federal and state agencies and departments, as well as for local governments and nongovernmental organizations. The plan is a comprehensive analysis of threats to the environment in the Keys and proposes more than 90 specific strategies to address these threats. It sets priorities, commits agencies to monitoring programs, and defines research for updating and adjusting the plan (NOAA, 1996).

Despite the success of the planning effort, NOAA will face significant challenges in the future. The Core Group that wrote the plan has been dissolved, even though success depends on a continuous planning process in the Keys. NOAA must make decisions about the composition and role of the Advisory Council and other planning bodies. Upcoming decisions about major issues (e.g., controlling the impact of water quality) depend on NOAA's ability to sponsor or conduct high-quality scientific research to resolve uncertainties about how the Keys ecosystem functions.

Mangrove Forest, Key Largo, Florida Keys National Marine Sanctuary. Photo courtesy of William Eichbaum.

Reasons for success or failure: After the final sanctuary management plan was issued (NOAA, 1996), a local referendum in Florida on the future of the sanctuary was held in November 1996. The result was 55 percent to 45 percent against the plan. Although the referendum was only advisory and NOAA plans to continue implementing the plan, the long-term outcome is not yet clear.[1] The management plan is at the cutting edge of widely discussed theories about how to move from narrow, single-purpose management systems to "community-based" environmental management systems.

In pursuit of its ambitious goals, the plan achieved some notable preliminary successes. The plan represents the first comprehensive effort to identify and assess the complex operation of the natural system of the Keys. The planning process was based on collaborative decision making, which required extensive interaction and cooperation among federal, state, and local agencies, as well as a range of nongovernmental constituencies. An important component in the collaborative process was the Sanctuary Advisory Council, an innovative group that included representatives of a wide range of local stakeholders. Unfortunately, the start-up of the council was delayed for several months for administrative reasons, and its

---

[1]In spite of the referendum, the governor of Florida and his cabinet unanimously approved the management plan on January 28, 1997. Approval was predicated on a resolution asserting the traditional authority of the state of Florida over state waters and agreeing to joint federal and state management.

role was not clearly defined. Nevertheless, despite, or perhaps because of this, the council eventually played a central role in the planning process.

Between July 1993 and March 1995, the plan disappeared from public view while details were negotiated and interagency reviews were completed. This lengthy delay resulted in a loss of momentum, and by the time the plan was finally published, opposition had been mobilized. In spite of the length of time spent preparing the plan (68 months), uncertainties and suspicions about NOAA's authority and intentions generated intense conflict.

The plan has yielded some benefits, however. New rules and better enforcement now protect the reef. Scientific uncertainties about the causes of water quality problems have been addressed. Elements of the plan have appeared in the county's recent land use plan, and data generated in the planning process have been used by state agencies to make decisions affecting the Keys. The structured planning process has been able to overcome fragmentation. Important elements of the planning process were a consistent core group of knowledgeable decision makers, partnerships with nongovernmental agencies, and a broad-based advisory council. It remains to be seen how effective this collaborative effort will be in overcoming conflicts among supporters and opponents of the sanctuary.

Key features: Influential legislators led the efforts to designate the sanctuary, and the legislation they helped enact went beyond previous legislation in important ways. It required NOAA to consider the full range of environmental issues and to create an advisory council. It required the EPA to work with the state to conduct research into the causes of water quality problems. Finally, it directed that zoning be considered to restrict certain activities. These broad mandates prompted NOAA to involve professional planners from its Strategic Environmental Assessment Division in the planning process. They facilitated activities of the core group, which included representatives of federal, state, and local agencies. Both the core group and the advisory council were new kinds of organizations within the sanctuary program, and NOAA had to learn through experience how to coordinate their activities.

The plan was developed in the context of a long history of struggle between county and state governments over land use in the Florida Keys. Environmental issues and whether or not to impose controls on development are central to local politics. Not surprisingly, the draft management plan provoked intense controversy over the specifics, especially zoning, and over the power of the federal agency (NOAA) in general.

Enabling factors: There was general consensus among stakeholders that the Florida Keys are unique and that everyone had an interest in preserving the unique character of the area, despite disagreements about how the Keys should be managed. The legislation itself was extremely broad, compelling NOAA to take novel approaches to sanctuary planning. These approaches were supported by NOAA's preexisting relationships with state agencies and by the existence of two smaller sanctuaries in the Keys. Shortly after the legislation was passed, NOAA officials and the governor's office agreed to work cooperatively and informally to develop the management plan. The manager of Looe Key NMS became the manager of the entire Florida Keys NMS, and his network of relationships, his personal style, and his ability to foster communication and collaboration were important elements in the planning process.

The core group and the advisory council adopted different approaches to decision making. The core group was much more formal and organized than the advisory council, which was more informal and people-oriented. As a result, problems and constituencies were approached in a variety of ways, which increased the chances of success. Finally, Florida has much more authority to intercede in local land use decisions than many other states. Thus, the state was able to use the results of the sanctuary planning process to make management decisions to protect the environment of the Keys.

Southern California Outer Coastal Shelf Oil and Gas Leasing

Oil development in the Santa Barbara Channel, northwest of Los Angeles, has a long history. The nation's first offshore oil well was drilled in this area in 1898 from a pier extending from shore. Santa Barbara County passed moratoria on offshore wells as early as 1927. Recent governance efforts must, therefore, be seen in the context of a long-standing conflict between constituencies for whom resource extraction is paramount and constituencies for whom conservation is

Oil and Gas Platform located offshore, Carpinteria, California. Photo courtesy of the Minerals Management Service, U.S. Department of the Interior.

paramount. In the last 20 years, there have also been rapid changes in the governance system for marine areas in California. During this period, California's Coastal Act was enacted, and consistency review (to promote consistency between state and federal policies as required by the CZMA) was extended to the state. During this same period, local, regional, and state authorities enacted increasingly strict controls on air pollution from a wide range of sources, ultimately including activities on the OCS. At the same time, federal agencies, such as MMS, have tried to develop more flexible policies that are responsive to local circumstances and to streamline their decision making. MMS' efforts to adapt to the changing circumstances in southern California has resulted in the creation of novel collaborative processes, the resolution of several long-term disputes, and an increase in oil production.

Against this background, Exxon installed three offshore oil and gas drilling platforms in the Santa Ynez unit, the first in 1976 and two more between 1989 and 1992. In addition, Exxon's on-shore processing plant at Las Flores Canyon began operation in 1993.

Reasons for success or failure: In response to demands that local and state governments play a larger role in making decisions that affected them, the MMS learned to share authority with communities and governments. The policies adopted by MMS, the state, the counties, and the oil companies as a result reduced overt conflict and friction and increased the system's ability to adjust to changing conditions. Permitting processes were streamlined, and an integrated regional planning approach was adopted. This approach, in turn, produced better environmental reviews to support decision making and fostered more productive, long-term working relationships among the parties. Significant agreements were negotiated regarding air pollution and oil transportation and processing, which resulted in the licensing of additional facilities and increased production and a net improvement in regional air quality.

A significant indicator of success was a recent statement by the Santa Barbara County Board of Supervisors that, with the proper safeguards, they might consider allowing additional leasing in the region. However, federal restrictions on the involvement of nonfederal employees in contracting decisions resulted in lengthy delays and eventually undermined trust in federal responsiveness to local requirements.

Key features: Historically, local opposition to oil development has been strong. Santa Barbara County had passed moratoria on offshore wells as early as 1927 for aesthetic reasons. It is no surprise that more recent development provoked intense political battles over air pollution, facility siting/licensing, and transportation. For example, in 1976, Exxon responded to local attempts to control air pollution from OCS activities by anchoring an offshore storage and treatment facility just beyond the area of state and local jurisdiction, in federal waters, an act that was seen by many local stakeholders as arrogant disregard for state and local concerns. At the time, neither the state nor the county had jurisdiction or leverage over OCS activities, and officials at each level of government came to see the facility as a highly visible symbol of governance problems relating to oil and gas development.

In this context, MMS, realizing that changes had to be made to come to grips with regional and local conditions, loosened the links between management policies in the Gulf of Mexico and in California. MMS also agreed to share significant authority with other parties, resulting in the creation of well defined, collaborative, decision-making mechanisms. This led, for example, to the establishment of joint review panels, which produced a single document that satisfied a variety of federal and state environmental requirements. MMS officials now regard the joint review panels as part of their new way of doing business.

Over a period of years from the late 1970s through the 1980s, Exxon proposed a variety of development scenarios, including offshore and onshore pipelines, a marine terminal, using shuttle tankers to deliver oil, the offshore storage and treatment facility, and an onshore processing plant. This range of options opened several doors for negotiation among local, state, and federal agencies. At critical junctures, both Exxon and Santa Barbara County turned to allies in the federal government for support; however, their inability to win a clear victory at the federal level forced them to return to the regional negotiating arena.

By the end of 1985, all of the parties (county, state, Exxon, Department of the Interior) were entangled in lawsuits about air quality, oil transportation, and future development. Without necessarily conceding their major points, the parties negotiated a series of agreements that resolved these issues and set the stage for building additional platforms and an onshore processing facility. Once the long-standing issues were out of the way, MMS, the county, and the oil companies in 1993 initiated a study to address the issue of phasing in future development. MMS' efforts in southern California thus represent a significant break with its traditional way of treating OCS development without the involvement of local coastal communities.

Enabling factors: The blowout and oil spill of 1969 sensitized local residents and governments to the possible consequences of oil development and aroused fierce opposition to further development. The 1976 amendments to the CZMA gave the state the power to review federal actions in the OCS for consistency with the state's coastal plan. The importance of air pollution in southern California provided state and local agencies with additional leverage over offshore development, particularly after the Clean Air Act Amendments of 1990 gave EPA control, which EPA delegated to the county Air Pollution Control District. A series of moratoria on the sale of OCS leases, beginning in 1982, added to oil company incentives to increase production on the leases they already held.

In this environment, a new MMS regional administrator was appointed who recognized that local residents and governments had to be included in decision making. His personality and character traits were vitally important in establishing and maintaining relationships among the parties. In addition, the new administrator initiated and/or agreed to specific policies, including collaborative power-sharing processes, conflict resolution training for MMS staff, and moving the office from Los Angeles to Ventura County, all of which contributed to a useful framework for problem solving. This flexibility was supported by MMS headquarters, which allowed the local office considerable latitude. Finally, after a long history of conflict, most participants had come to believe that collaboration is better than fighting each other in the courts.

Gulf of Maine Fisheries

The Gulf of Maine is the largest semi-enclosed shelf sea bordering the continental United States (Christensen et al., 1992) and includes almost every conceivable use of the marine environment. The gulf watershed extends from Nova Scotia and the Bay of Fundy in the northeast to Boston and Cape Cod in the southwest. Nearly one-third of the gulf's relatively sparsely populated coastline is estuarine habitat. The area offshore, notably on Georges Bank, over the western Nova

Stern of gillnetter boat with crewman dressing groundfish off the coast of Massachusetts. Reprinted with permission from Commercial Fisheries News, *copyright Compass Publications, Inc. Photograph by Richard Burnham.*

Scotia shelf, and in the Bay of Fundy, has been extremely productive and historically rich in lobster, scallops, groundfish, and other stocks.

Because of the low population density and the small amount of river runoff
(and hence small pollutant loads), the marine waters of the Gulf of Maine do not
show the serious signs of environmental degradation typical of more urbanized
regions (Waterman, 1995). However, cumulative detrimental impacts on habitats
and on the living marine resources are evident in many harbors and estuaries
throughout the Gulf of Maine (NRC, 1995c). In particular, intensive harvesting
of groundfish stocks, especially cod, haddock, and yellowtail flounder, has led to
severe declines in these species. Most of the commercially important stocks have
fallen below or are approaching the lowest estimated levels on record. Two stocks,
the Gulf of Maine haddock and the Georges Bank yellowtail flounder, have been
declared commercially extinct (NRDC, 1997). The overall levels of groundfish
are extremely low, as is the estimated spawning biomass for these species. Chemical contamination of fishery resources has led to several fishery advisories or
closures (Capuzzo, 1995).

Stock declines have resulted in substantial economic losses and the social
disruption of fishing communities. The New England Fishery Management Council has closed fisheries and issued moratoria in an attempt to rebuild stocks, and
both federal and state governments have implemented various support plans to
lessen economic impacts, but the future of the offshore fishing industry in this
region is still in doubt.

Reasons for success or failure: New England groundfish stocks are in a state
of collapse. There is widespread agreement among the scientific community that
this collapse was caused by overfishing (Hoagland et al., 1996). As Murawski
(1996) summarizes:

> Groundfish...have not fared well under domestic management....The rapid in-
> crease in fishing effort during the late 1970s and early 1980s resulted in in-
> creased fishing mortality rates....In recent years, fishing mortality rates have ex-
> ceeded recruitment overfishing levels by a factor of 2 or more....The recent
> decline in these offshore resources is attributable to persistent, gross recruitment
> overfishing...declines in stock sizes and landings could have been averted or at
> least mitigated if the stocks had not been significantly recruitment overfished.

Key features: Offshore fisheries in the Gulf of Maine are managed under the
provisions of the Magnuson-Stevens Fishery Conservation and Management Act
(FCMA) of 1976. This law was passed largely in response to concerns that foreign fleets were depleting fish stocks located near the United States. In fact, New
England fishermen sailed up the Potomac to Washington, D.C., in 1974 to protest
the impacts of foreign fleets. The New England Fishery Management Council,
under the FCMA, manages fisheries in the Gulf of Maine and includes voting
representatives of the fishing industry, the NMFS, and coastal states in the region.
The council is responsible for enacting fishery management plans for each fishery

in its jurisdiction in accordance with the statutory principles of the act. These include preventing overfishing while maintaining optimum yield, basing decisions on the best scientific information available, and promoting efficiency in the utilization of fishery resources.

Before 1971, restrictions on mesh size were the preferred approach to prevent overfishing under an international management regime. Beginning in 1971, total allowable catches were set on the major groundfish species; in 1974, quotas for each fishing nation were established. Since 1976, the New England council has implemented a variety of management regimes in attempting to meet these goals. In general, the management systems have oscillated between input and output controls. Input controls include gear restrictions (e.g., minimum mesh size for trawl nets) or the prohibition of specific technologies. Output controls, such as total allowable catches (TACs), limit the number of fish that can be harvested. By 1982, because fish populations had not recovered, the quota-based system was replaced with a system of input controls: minimum mesh sizes and minimum fish sizes. These were tightened periodically until 1994, when the council placed a moratorium on new entrants to the fishery, limited the number of allowable days at sea, and restricted the extent of fishing (Amendment 7). Finally, an amendment to the fishery management plan adopted in 1996 established a TAC that, when exceeded, triggers further restrictions.

Enabling factors: The various shifts in the management regime were attempts to respond to problems that arose during each previous regime. Thus, individual nation quotas were adopted in 1974 after the haddock stock had collapsed from pulse-fishing with small mesh nets by the Soviet trawler fleet in the 1960s. However, quotas led to "derby" behavior (also called the "race for fish"), causing the fishery to be closed down earlier each year as more vessels entered the open-access fishery. This system was later modified to limit the catch on individual trips, but it was impossible to monitor daily landings of all vessels. Catches were often mislabelled and/or landed illegally, false fishing locations were reported, and other forms of noncompliance increased. The stock recovery in the 1970s prompted a return to gear-based (or input) regulation in 1982. However, the open-access fishery and the lack of an effective TAC program contributed to another cycle of severe overfishing. This is an example of a phenomenon described as "the solution becomes the problem" (Checkland 1982; Clemson 1984).

Other factors were also significant in this management history. When the FCMA was passed in 1976, the New England fishing fleet was composed predominantly of small, old vessels that concentrated mainly on near-shore fishing. The open-access policies of that period, combined with ample federal subsidies and support programs, provided substantial incentives for fleets to expand, which ultimately led to overcapacity. In addition, the New England council had difficulty developing and applying an operational definition of overfishing. Although regulatory guidelines mandated the development of a definition, the council was not given a specific deadline. This allowed disagreements to persist within the council about the status of individual stocks and delayed consensus about overfishing and the implementation of a stock rebuilding program. Finally, the Conservation Law Foundation brought suit against NMFS, and deadlines were established as part of the consent decree settling the suit.

The governance regime was unable to resolve the fundamental tension among the FCMA's statutory goals, which called for stock conservation, optimal yields, and economic efficiency. As a result, valid scientific warnings of imminent stock collapses were not acted on. NMFS and the council were subject to political pressure by interest groups acting outside of the formal governance structure (the "end run" phenomenon).

Alaska Fisheries By-Catch[2]

In the fall of 1995, the Alaska Fisheries Development Foundation, with funding from NMFS, conducted a series of workshops in Alaska to explore solutions to by-catch problems in a wide range of commercial fisheries. Workshop participants included key decision makers in the industry, including the heads of major professional and industry associations, executives of large fishing firms and fish processing companies, government fisheries scientists, representatives of Native American groups, the director of the regional fisheries management council, and the local representative of an international environmental organization.

Reasons for success or failure: Although the workshops did not result in the implementation of solutions, they represented a notable break with the past. They provided tangible evidence of the opportunity for and the benefits of collaborative problem solving. By starting to bridge gaps not only between regulators and the fishing industry but also among different groups within the fishing industry, the workshops raised hopes that by-catch problems could be addressed locally.

Many of the participants remarked that this was the first time they had met to discuss common problems outside the formally regulated (and often divisive and competitive) stock allocation process. In this new forum for problem solving, participants found common ground they had not previously been aware of, regulators received valuable feedback about impediments to problem solving, and novel solutions were proposed and discussed from a wide range of viewpoints. Participants came away with tangible experience of constructive cooperation. However, because no permanent infrastructure was established for collaborative decision making, no further progress on resolving by-catch problems has been made.

Key features: The workshops were characterized by collaboration, active participation by key decision makers, and a format that established equality among

---

[2]By-catch refers to the indicental take of nontarget species in fishing operations.

all participants. The workshops were place-based in that they focused on a geographically defined area. They constituted a new forum outside traditional decision-making processes.

Enabling factors: The severe conflicts in the past had demonstrated the potential dangers of noncooperation. Growing awareness of the potential repercussions of continuing by-catch problems provided an additional impetus to cooperation. Frustrations that local problem-solving was stymied by the existing regulatory structure had been building. Hands-off funding support from the NMFS was the catalyst for change, and allowing local decision makers to develop their own ideas, facilitated by uninvolved outsiders, created new channels of communication.

Maine Lobster Fishery

Throughout its history, the seasonal Maine lobster fishery has been dominated by independent fishermen using boats under 36 feet in length. Until recently, the catch was effectively (though indirectly) controlled by the inefficiencies of wooden traps and limitations on navigational technology, sounding gear, and boat speed and size. The number of fishermen was also limited because the rules required that fishermen live close to the area they fished, encouraging a sense of stewardship among fishermen, who identified personally with specific areas. At the same time, it allowed fishermen a level of control over who fished and how lobsters were harvested.

Recently, new technology has upset these relationships. More rugged and efficient wire traps, larger and faster boats, and more sophisticated navigational gear have enabled fishermen to fish much larger areas and to fish them much more intensively. As a result, both the number of fisherman and the size of the catch have increased rapidly in the past decade. To control expansion and limit damage to lobster stocks, the state of Maine passed fishermen-supported legislation to create a democratic, bottom-up management process focused around local fishery councils. This process incorporates controls (i.e., limits on the number of traps per boat) and an apprenticeship plan to limit access to the fishing grounds.

Reasons for success or failure: This plan appears to have the potential to prevent the overuse of resources and promote sustainability. It has provided a democratic mechanism for resolving conflicts and implementing solutions to access and enforcement issues. However, the plan does not explicitly take into account scientific knowledge about the dynamics of the lobster population. In addition, it is too early to tell if the plan will be successful (it has only been in place since mid-1996).

Key features: The overuse of resources has been avoided by protecting both juveniles and the older brood stock by imposing size limits and by removing egg-

(A) Setting lobster traps. (B) Skipper and sternman picking lobster traps. Reprinted with permission from Commercial Fisheries News, *copyright* Compass Publications, Inc. *Photographs by Sarah Sherman Brewer.*

bearing adults from the harvest by requiring that they be marked when caught. These conservation measures have been retained, at the insistence of the fishermen, despite efforts by the state legislature to remove them. Fishermen are protected from excessive or destructive competition by limitations on the number and size of traps per boat. Even though a statewide limit was established by the legislature, local management councils have the option of voting to establish lower limits as well as modifying or establishing other controls in each district. An apprenticeship program limits fishermen to those who have completed a two-year training program, thus mitigating the "open-access" problem that contributed to the decline of the offshore groundfish stocks in the Gulf of Maine. Controlling both technology and the size of the harvesting unit (trap limits) rewards skill and encourages local efforts to husband wild stocks. This structure is based on the delegation by NMFS of certain management authority to the state, and then by the state to local, democratic management councils, whose conduct reflects their local self-interest in maintaining a healthy stock.

Enabling factors: There was widespread recognition among fishermen that more advanced technology and lower barriers to entry constituted a serious threat to the fishery. In a series of opinion surveys, more than 85 percent of fishermen expressed a desire for trap limits. There was a strong tradition of local resource control and decision making, and the state was committed to a collaborative approach as the basis for improving the management regime. Not only was NMFS willing to share power with the state, but the state was also willing to share power with local management bodies. Finally, there was a sufficient base of scientific knowledge, as well as a desire to improve the relationship between the scientific and fishing communities.

Chesapeake Bay Program

Efforts to restore and protect Chesapeake Bay began in the late 1970s when legislation directed the EPA to carry out a study of the problems of the bay and recommend solutions (Eichbaum, 1984). The study was jointly directed by EPA and the states of Maryland, Virginia, Pennsylvania, and the District of Columbia and involved many scientists who had studied the region. The study produced integrated recommendations for strategies to correct the problems of the bay. These recommendations covered a wide range of issues, including point and nonpoint source pollution, monitoring and evaluation, and future management structures. The several states embraced most of the recommendations and added suggestions for land development practices and resource management to address problems associated with oysters and striped bass (U.S. Department of Natural Resources, 1995).

Beginning in 1984, each state, the District of Columbia, and the federal government began to develop and implement programs to restore the Chesapeake

Striped bass restored in Chesapeake Bay after strict management regime. Photo courtesy of William Eichbaum.

Bay. A unique aspect of this management program was the joint system of governance, which was created voluntarily by the several jurisdictions. The essence of this system was an executive council that included the state governors, the mayor of the District of Columbia, and the EPA administrator. Under the executive council, a management committee was given more routine responsibilities for overseeing the implementation of specific programs. Working committees ensured that proper attention was given to specific issues, such as monitoring and resource management. This structure was duplicated in parallel scientific and citizen advisory committees.

Reasons for success or failure: The model developed for restoring the Chesapeake Bay has proven to be very successful. It has been a resilient vehicle for more than a decade of regional management of the bay and has served as a model for estuary governance across the nation. The success is attributable, in large part, to the political leadership that has been exercised from time to time by various members of the executive council.

Key features: The intimate involvement of regional scientific experts, interested citizens, and other economic stakeholders has been important to success. Participation was voluntary, and the structure was tailored to the technical as well as political traditions of the bay region. An important element of political accountability has been an intensive monitoring program, which allows all interested parties to track accomplishments. In addition, the executive council has provided a public platform for making bold commitments to good governance, such as the 1987 decision to reduce the discharge of nutrients into the bay by 40 percent (EPA, 1995).

Enabling factors: This governance structure was possible because political leaders responded to a dramatic threat to a highly valued resource. There were no appropriate models for voluntary approaches to governing such a large region or such a complex set of issues, so the situation required experimentation in the context of political commitment. However, as time passes and some indicators of successful restoration appear, it is becoming less certain that this voluntary system will continue to work as effectively, especially for the jurisdictions with the least to gain from protecting the bay.

Long Island Sound National Estuary Program

Long Island Sound is one of the country's most urbanized estuaries. In the late 1980s, despite improved wastewater treatment in the New York metropolitan area and Connecticut, water quality in the sound deteriorated. Particularly notable was extreme hypoxia (low dissolved oxygen) during the summer months. In an effort to address these problems, Long Island Sound was designated for inclusion in the NEP (National Estuary Program) in 1989.

The NEP planning process involved extensive analysis of the water quality, including the development of a complex computerized model of water circulation within the sound. These studies revealed that the primary water quality problem was an excess of nitrogen, which stimulated algae blooms that subsequently died and created the low oxygen level. The Policy Committee of the NEP, including representatives of the states of New York and Connecticut, adopted a "no net increase of nitrogen" policy for wastewater discharges from the two states.

This policy was administered on a regional basis, requiring that treatment plants work cooperatively. Treatment innovation at the local level was necessary to meet the overall performance standards for dischargers. There have already been significant reductions in nitrogen levels in both states, and more are expected. The final CCMP (Comprehensive Conservation and Management Plan) for the NEP used the computer model to refine nitrogen reduction targets by revealing where the greatest environmental benefits could be achieved at the lowest cost.

Reasons for success or failure: Real reductions in nitrogen loadings have been made and are expected to continue. Support for the plan to improve water quality in the sound is widespread. A means has been established for making the most cost-effective investments first and then for evaluating the results before deciding on further, more costly, reductions.

Key features: A commitment was made both to broaden public participation in decision making and to evaluate key problems scientifically. The "no net increase" in nitrogen policy and the subsequent CCMP focus on results and allowed individual treatment plants, local governments, and states to find technical solutions that fit their needs. The freedom to innovate to achieve stated goals was a central feature of this program.

Enabling factors: Funding was sufficient to complete the computer model, which was an essential evaluation tool. Combined with other scientific information, this model provided a basis for decision making. The working relationship between the states and EPA was excellent, and constituent groups provided strong support for positive action by the NEP Policy Committee.

Oregon Rocky Shore

Oregon established a planning program for ocean resources in 1987 in response to federal proposals for offshore oil and gas and marine mineral exploration. The legislature amended the program in 1991 to create an Ocean Policy Advisory Council in the Office of the Governor, staffed by the Oregon Coastal Management Program (CMP). The council's initial focus was on the degradation of rocky shore habitats caused by a variety of human activities.

A strategy was developed with an ecosystem-management approach involving all relevant state and federal agencies as well as the public. The strategy was supported by information from a site inventory by state and federal resources agencies through a special four-year grant from NOAA to enhance the Oregon CMP. The Rocky Shores Strategy, now part of the Oregon Territorial Sea Plan adopted as part of the Oregon CMP, contains overall goals and policies and addresses the natural resources and management issues of each site, including rocky cliffs, rocky intertidal sites and associated submerged rocks and reefs, and offshore rocks and reefs. The management strategy was accompanied by a campaign to increase public awareness and foster a sense of stewardship of rocky shore resources. Several state agencies have begun to implement the strategy through a combination of regulations, on-site programs, additional planning, and work with local volunteer groups.

Reasons for success or failure: The rocky shores strategy has provided the first-ever comprehensive assessment and management plan for all rocky shores in the state. It has focused the attention of two key state agencies on the many issues confronting management of Oregon's rocky shores and has provided a framework for resolving the most serious conflicts between resource use and resource conservation. The program is successful because it is based on sound information, involves all relevant agencies, responds to site-specific situations but within an ecosystem context, and involves the public.

Key Features: A detailed rocky shores inventory and resource analysis was conducted over a period of two years by the Oregon Department of Fish and Wildlife in cooperation with the U.S. Fish and Wildlife Service and the University of Oregon Institute of Marine Biology. Additional resource inventories and site surveys were conducted on offshore kelp reef areas. An open planning process allowed various interested parties to help fashion the goals, policies, and site-management prescriptions and gave a much wider than usual range of stakeholders a voice in the outcome of the planning process. A communications strategy, linked directly to overall goals and objectives, as well as to needs or opportunities of each site, was developed by an interagency group and is now being implemented at selected sites.

Enabling Factors: Public support was crucial to this project. In fact, a series of public workshops and listening sessions at the beginning of the planning process revealed a high level of public concern about the degradation of rocky shores. Funding from NOAA enabled the state to acquire necessary inventories and to perform analyses so that specific management prescriptions could be fashioned to meet on-site needs. The existence of the state program, with its own clear policies and coordinated means of enacting them, enabled the state to play a strong role in determining the rules of engagement with federal ocean resource agencies. As a result, state managers were able to ensure that the program was as responsive as possible to local needs and circumstances. The evident effectiveness of the strategy led several key state and federal agencies to embrace it as a means of resolving long-standing management problems.

San Francisco Bay Demonstration Project

The San Francisco Bay Demonstration Project represents one of several attempts by NOAA's National Ocean Service (NOS) to develop new ways of refining its products and services, more effective ways of targeting local and regional needs, and to improve the efficiency of its operations. In San Francisco, NOS attempted to find ways to package its technical services and data to support shipping, port management, and coastal resource management. It did so in large part through a series of discussions with local agencies and groups to identify key local issues and information needs. The staff of the San Francisco Bay Demonstration Project then worked with these groups to find ways to repackage the data NOS gathers to meet local needs.

Reasons for success or failure: The San Francisco Bay Demonstration Project focused on providing tangible help for meeting specific needs, mostly by finding ways to overcome information fragmentation and the absence of data in these areas, thereby enhancing opportunities for realizing economic benefits from coastal resources. For example, more accurate and readily available bathymetry, navigation tools, and water level information made it possible for the Port of

Cargo-handling facilities, Port of Oakland, California. Photo courtesy of William Eichbaum.

Oakland and shipping companies to operate more efficiently by increasing the effective depth of existing navigation channels. By acting as a service agency, and by emphasizing cooperative solutions, NOS helped build new working relationships among agencies in the region. Future conflicts over competing uses may be reduced by making better information more readily available to stakeholders.

Key features: The project reached out to local stakeholders to identify local issues and needs and focused on building relationships rather than on dictating solutions. Potential solutions were defined through a collaborative process that emphasized local involvement. The project had a clearly defined overall goal of making better use of available scientific information and was clearly place-based since it identified San Francisco Bay as its service area.

Enabling factors: The San Francisco Bay Project was a conscious attempt by a federal agency to play a new role. The attempt was supported by a clearly stated policy to that effect. Success in fulfilling this role largely depends on the ability of NOS's local representatives to improvise and behave opportunistically by establishing relationships wherever local agencies are receptive to the idea rather than by attempting to follow a rigidly prescribed plan. The fact that NOS had valuable information increased the chances of success. Finally, the NOS site manager had the necessary negotiating, marketing, and leadership skills to get the job done.

Santa Monica Bay Restoration Project Regional Monitoring

The Santa Monica Bay Restoration Project (SMBRP) is a part of the EPA's NEP. Its primary charge is to develop and implement a regional management plan to protect and restore resources in the region. Working in close association with a regional regulatory agency (the Los Angeles Regional Water Quality Control Board) and through a series of committees with broad stakeholder participation, the SMBRP developed a Comprehensive Conservation and Management Plan to provide regionally integrated information about compliance, risks to human health, the status of regional resources, and the success of restoration. The SMBRP held a series of planning workshops to prioritize issues and establish ground rules for revising existing monitoring programs. The SMBRP then empaneled a series of working groups to draft and implement revisions to the monitoring program.

Reason for success or failure: The working groups created regionally integrated programs from existing fragmented monitoring programs without raising costs and targeted these revised programs to meet the information needs of specific decision makers. As a result, these programs adopted different strategies (e.g., day-to-day empirical health management, more formal longer-term risk assessment) appropriate to different issues. By focusing rigorously on how monitoring information would be used, the work groups used the program design process to streamline decision-making relationships at the agency level, clarify policy, and reduce conflicts among stakeholders. The revised programs standardized assumptions and methods and specified information flow. Because all key stakeholders participated, the working groups strengthened cooperative relationships among agencies.

Key features: The NEP provided long-term goals that were clear enough to focus efforts but flexible enough to accommodate local and regional variations. Thus the NEP furnished leadership and guidance without exercising undue control. The Los Angeles Regional Water Quality Control Board also adopted a new role. They set boundary conditions and facilitated and managed the process rather than mandating solutions. The working groups followed ground rules developed in previous planning steps. These included an implicit cap on costs, a commitment to collaborative and consensus-based decision making, an understanding that program revisions had to be based on sound science, and a focus on improving coordination. All relevant stakeholders were included, even if they had not historically been involved in compliance monitoring programs. Positive momentum was created and maintained by incrementally implementing revisions to existing programs as soon as they were agreed on by all participants. Shared decision making validated each participant's interest, and the geographical boundaries of the project helped maintain a common focus.

Enabling factors: Pre-existing and long-term professional and personal relationships among many of the stakeholders contributed to the feeling of trust and the sense that the SMBRP's activities were, in part, a continuation of past work. As a result of this shared history, there was a tacit commitment to collaboration and an informal network of back channels for communication and negotiation that eased the process of consensus building. In addition, there was a long history of interest in the region, especially among dischargers, in improving the monitoring system. A study of monitoring in southern California (NRC, 1990b) had provided an unbiased forum for "truth telling" about the flaws in the system and had led to a widespread (and public) acceptance of the need for change. A large existing body of scientific information and empirical knowledge about the ecological system provided a firm basis for proposing revisions.

SUMMARY OF THEMES

Certain common themes have emerged from these examples of existing coastal and marine programs and activities:

- Every situation is different in terms of historical background, initiating events, leadership style and capacity, scientific issues, natural resources, organizational structures and flexibility, and the availability of scientific information.
- Key initiating events, even if they seem negative (e.g., oil spills, tanker groundings, severe conflicts), can catalyze useful changes in behavior.
- Focusing on a specific place or geographic area is necessary for creating a sense of shared purpose.
- Valid, relevant scientific information can provide a sound basis for decision making.
- The processes, even the successful ones, are chaotic and unpredictable, and favorable outcomes sometimes depend partly on sheer luck.
- Skilled leadership is a necessity.
- Successful leadership and decision making are characterized by a high degree of improvisation.
- Organizational and behavioral issues are as important as scientific issues (often more so).
- Collaborative processes are necessary to, but not sufficient for, success.
- New, more flexible roles for federal agencies make novel solutions possible.
- New forums for discussion and decision making can foster creativity.
- Patterns of power sharing evolve.

Only a few of these common themes, particularly the need for valid scientific information and the importance of flexibility, correspond to the principles developed by the committee at the outset of the study (see Chapter 1). There are three reasons for this. First, the principles were developed in the abstract, without reference to specific instances of governance. Second, the principles refer in large part to final outcomes (e.g., improved economic efficiency, sustainable development) that are desirable from the perspective of society at large. In contrast, the common themes from the examination of existing programs and activities refer to more immediate, process-related issues of initiating and successfully sustaining and managing specific governance programs. Third, the principles implicitly pertain to larger-scale, more complete governance systems that have the ability to create these final outcomes. The common themes from the review of examples reflect the more fragmentary, piecemeal nature of real-world governance, which often focuses on a portion of the overall governance process.

Two fundamental lessons emerge from a comparison of the initial principles with the common themes drawn from real-world examples:

- Opportunities for improving governance come in all sizes and shapes. Some are large-scale efforts that can influence final outcomes; others are smaller-scale ventures with more limited goals.
- Organizational and behavioral issues are vital to every project, no matter what the scale.

In general, then, the common themes focus on process-related issues that are essential delivering the positive outcomes identified in the initial principles. Therefore, the committee determined that organizational issues should be further examined.

4

Organizational and Behavioral Issues

The following discussion extends the examination of lessons learned from the case studies and other examples and describes the organizational and behavioral issues the committee determined to be essential to successful marine area governance and management.

- Marine area governance problems are complex but have certain predictable characteristics.
- Complex problems necessarily involve a range of nonscientific issues having to do with communication, human behavior and motivation, and how people organize to deal with them.
- Traditional hierarchical, bureaucratic approaches to problem solving are not appropriate to these types of situations.
- Recent research into organizational structures and functions has much to offer, but no single organizational structure will solve all problems.
- Guiding principles rather than organizational prescriptions can provide direction for problem solving in this arena.

COMPLEXITY OF GOVERNANCE PROBLEMS

The committee's three in-depth case studies and other examples of marine governance and management programs included such diverse issues as living resource management (Gulf of Maine), regional planning (Florida Keys), environmental monitoring (Santa Monica Bay Restoration Project), and the distribution and sharing of information (San Francisco Bay Demonstration Project). Virtually all of these situations were different in terms of key characteristics, such as historical background, initiating events, existing institutional framework(s), the roles played by federal agencies, local leadership style and capacity, scientific issues, and the availability and thoroughness of scientific information. The variety of problems and factors complicated the committee's attempt to establish guidelines for effective marine area governance.

Despite their differences, the fundamental attributes of all of the situations were typical of complex policy and management problems. In general, these include:

- multiple aspects that interact in complex, often unpredictable, ways
- no simple, easily achievable solutions
- scientific uncertainties
- a large number of participants (or stakeholders) with different, often conflicting, priorities and perspectives
- either active conflicts or the residue of past conflicts
- competing claims for leadership and/or authority
- nonexistent, confusing, inappropriate, or overlapping regulatory and management mechanisms
- fluid, poorly defined, and unstable decision-making processes

Clearly, situations with these attributes are poorly suited to the classic model of decision making (variously called "formal," "instrumental," or "bounded" rationality), in which the advantages and disadvantages of well defined alternatives are analyzed and a decision is made by choosing one of them (Simon, 1957; March and Simon, 1958; Kalberg, 1980; Zey, 1992). Participants in several of the examples (e.g., Chesapeake Bay, Florida Keys, Oregon Rocky Shore, San Francisco Bay Demonstration Project, Santa Monica Bay Restoration Project, Southern California OCS) stated that, in effect, they had had to invent the process as they went along. This was often true even when general programmatic guidelines were available (for the NMS and NEP). Thus, although many examples of successful governance could be described in hindsight as if they had been carefully planned, in fact they involved confusing, often chaotic, processes that required intuition, improvisation, and downright luck. This situation has also been observed in recent research on managerial decision making (Isenberg, 1985), which suggests that high-level managers faced with complex problems depend more on intuition than on formal, analytical processes. In fact, both the MMS regional director in southern California and the manager of the Florida Keys NMS, stressed the role "gut" feelings, flexibility, and improvisation played in their respective situations.

Recognition of the complex nature of problems has contributed to fundamental changes in organizational theory and practice (Janowitz, 1959; Clark, 1985; Weick, 1985; Heydebrand, 1989; Miles and Snow, 1992). Rather than hierarchical control and a strict division of labor and responsibility, the new paradigm emphasizes a plurality of hierarchies that respond to a shifting network of mutual constraints and interactive influences. Rather than viewing complex systems as aggregations of simpler units, components of complex systems are considered interdependent, with dynamic and idiosyncratic properties of their own and with a high potential for emergent properties.

This paradigm emphasizes the importance of adaptation to changing and uncertain environments. The concept of adaptation, in turn, related the concepts of flexibility, changing boundaries among system parts, complexity, and surprise (Rochlin, 1989). Two of the case studies illustrate the distinction between the old and new approaches. In the Gulf of Maine, the formally defined catch allocation process allowed for participation by only a few stakeholders, prevented the adequate empowerment of scientists who had crucial information, and forced decisions to be made based on a restricted set of criteria. In contrast, the MMS in Southern California was free to create new decision forums and processes and to expand stakeholder involvement. As an aside, participants in the Alaska by-catch workshops were nearly unanimous in their conviction that the rigid, confrontational nature of the formal fisheries management process had prevented them from exploring creative solutions to industry-wide problems. The contrast between traditional and newer organizational paradigms has been summarized in terms of fundamental changes in the six characteristics listed in Box 4-1.

The shift basically involves moving away from centrally coordinated, rigidly structured hierarchies toward more flexible structures (Mintzberg, 1983; Miles and Snow, 1992; Wilson et al., 1994; Becker and Ostrom, 1995). The remainder of this chapter discusses various elements in this shift and their implications for marine area governance. Because the committee documented fundamental shortcomings in traditional governance systems, these elements are discussed in terms of their roles in promoting change.

Human Element in Governance Systems

Governance systems examined for this study were found to be far more than simple mechanisms for decision making. Rather, they were forums for exploring problems, pursuing and resolving conflicts, eliciting and establishing values, establishing power arrangements, and negotiating solutions, among other things. These characteristics were apparent in both large- and small-scale examples. In other words, governance systems necessarily operate both through organizational structures and through people. In general, the degree of success reflected the ability of a system to address and resolve the issues listed above. Thus, for example, the development of the sanctuary in the Florida Keys provided ample opportunity for the participants to propose, establish, and modify power arrangements and pathways for suggesting and negotiating solutions.

The intimate and unavoidable relationship between the process of governance and the organizational structures and mechanisms that embody it is complicated by the fact that organizations are dynamic systems that work not only to

BOX 4-1
Differences between Classical and Newer Organizations

Contrasting Perspectives	Classical Organizational Paradigm	New Organizational Paradigm
Simple vs. Complex	System is the sum of its parts.	Systems have emergent properties that cannot always be predicted.
Hierarchic vs. Heterarchic	There is a fixed hierarchy of problem components and organizational roles.	Hierarchies are not rigidly defined and can change.
Mechanical vs. Holographic	System elements act as predictable, single-action functions.	Elements constantly interact in complex ways.
Determinate vs. Indeterminate	System behavior is predictable.	Systems behave in ambiguous and uncertain ways.
Linear vs. Mutual Causality	Results come from simple, linear chains of cause and effect	Outcomes result from complex feedback and unclear/mutual causality.
Objective vs. Perspective	There is one "right" answer or point of view	The "right" answer varies depending on the point of view, values, and other factors.

Source: Adapted from Weick, 1985.

accomplish their stated mission(s) but also to maintain certain reward and punishment systems, power relationships, hierarchies, and so on. The relationship between behavior at the organizational and individual levels is necessarily close.

On the one hand, organizational structure and culture constrain, motivate, and direct individual behavior. On the other hand, individuals and their behavior embody the culture of the organization. Decision making, as well as attempts to change an organization's behavior, always results in losses and gains for its members on a wide range of axes (e.g., power, access to power, knowledge, status). Successful governance depends in large part on awareness of, sensitivity to, and management of this complex network of gains and losses.

Organizational Mission(s) and Goal(s)

Each organization (and each part of an organization) has one or more missions that justify its existence. Sometimes these are clearly stated, and sometimes they are implicit in the organization's past history. In addition, an organization can present one face to the outer world and a different face to its members. For example, the stated mission may be to provide a particular constituency with specific services while the true, internal mission is, in fact, to safeguard the political prospects of high-level managers. (This duality can be true of governmental and business organizations.)

Different combinations of implicit, explicit, external, and internal missions promote or constrain different kinds of behavior through a variety of mechanisms, including the organization's reward systems and its network of personal relationships and loyalties. Thus, changing the organization's core mission creates a great deal of tension because the mission is the rationale for making decisions and one means of resolving conflicts. In the Southern California study, it is clear that when the MMS regional director was promoting novel collaborative decision-making processes (i.e., changes in the agency's mission), he was perceived by the traditional MMS culture as a risk taker or "outlier."

A complicating factor is that a whole hierarchy or tangle of missions may be associated with different levels or parts of an organization and/or with a charismatic leader. Conflicts and contradictions among these missions is often a major source of organizational dysfunction and poor performance. For example, the management of groundfish in the Gulf of Maine was severely hampered by the inability to reconcile two competing objectives of fisheries policy—protecting the fisheries resource on the one hand and safeguarding the economic health of the fishing industry on the other. In this case, written mission statements were not enough to overcome the informal but extremely powerful forces that promoted the short-term interests of the industry at the expense of the long-term health of the resource.

Implicit and Explicit Reward Systems

Reward systems are behavior modification "carrots" and "sticks" that persuade and/or compel members of an organization to act in accordance with its culture. Reward systems usually operate on both explicit and implicit levels. Within organizations, explicit rewards are embodied in policies that define career tracks, pay increases and/or bonuses, and disciplinary measures up to and including termination. In governance systems that include multiple organizations, explicit rewards can include access to natural resources, permit approvals, or a seat at the decision-making table. Implicit rewards are always embodied in behavior, that is, what actually happens as opposed to what is theoretically supposed to happen. Implicit reward systems are almost always more powerful and more deeply rooted than explicit reward systems. Changing the ways an organization functions by addressing only explicit reward systems is often bedeviled by seemingly irrational behavior. In fact, people are merely responding to powerful, implicit rewards that remain in place and continue to exercise influence.

Examples of how reward systems work can be taken from the case studies. In the preceding discussion of organizational missions, the regional MMS director was described as being insulated from the dominant reward system, which might have inhibited risk taking. In the case of the Maine lobster fishery, the bottom-up management regime explicitly strengthened traditionally important implicit rewards (e.g., identification with a specific place, personal relationships among fishermen, a sense of stewardship). In the San Francisco Bay Demonstration Project, a redefinition of the goals provided the justification for rewarding different behavior that helped accomplish the new mission. In fact, changes in reward systems can be powerful supporting elements to changes in an organization's mission; conversely, leaving an old reward system in place can doom changes in the mission.

Existing Organizational Political Systems

Power is derived, allocated, and wielded in a variety of ways in organizations. Many organizations mimic political systems in society at large. In Davenport et al. (1992), organizational political systems are categorized as utopian, monarchical, feudal, federal, or anarchical. Utopian systems assume that behavior is rational and that participants will work selflessly toward shared goals. Monarchical systems are characterized by centralized power that is delegated only under tight control. In feudal systems, power is distributed among a group of "barons" who negotiate temporary alliances and power-sharing arrangements on an ad hoc basis. Federal systems allocate both centralized and distributed power in clearly defined ways, with the center coordinating rather than controlling power. Anarchical systems are free-for-alls in which power goes to whoever can seize it and keep it.

Categorizing the political systems in the various organizations the committee examined provides some insight into why they succeeded or failed. Several systems were predominantly federal in that the lead agencies guided, coordinated, and motivated participants. Federal systems included the Chesapeake Bay, the Florida Keys, Long Island Sound NEP, the Oregon shore, and the Southern California OCS. In these cases, the federal model was an important contributor to success.

The Santa Monica Bay Restoration Project provided an example of a combination of two political systems. When dealing with dischargers, over whom the Regional Water Quality Control Board had permit authority, the board set core goals and boundary conditions, while using a collaborative process to define a wide range of specific issues (a federal role). However, when the seafood tissue monitoring program required that the California EPA assess health risk, the board established an ad hoc arrangement in which the California EPA, as the end user of monitoring data, defined the fundamental conceptual approach while the regional board retained control over implementation (a feudal role).

Groundfish management in the Gulf of Maine can best be described as utopian. Its fundamental (and ultimately unrealistic) assumption that decisions would be based primarily on scientific information and its inability to account for and adapt to real-world behavior contributed to the failure of the fishery.

Changes in organizational structure or function must be made in the context of existing political systems. Transitions between some styles are more difficult than others, and changes are not usually reversible. For example, any attempt by MMS to recentralize decision making for offshore oil and gas development in southern California would undoubtedly encounter intense resistance from local governments and citizens who have grown accustomed to playing a role in decision making.

The type of political system is also closely related to reward systems, especially attitudes toward innovation and risk. Thus, utopian systems, which emphasize analysis, optimization, and technology, do not typically foster or support intuitive risk taking. This may explain why no innovative, "out of the box" steps were taken to rescue the fisheries in the Gulf of Maine. Finally, changes from one type of system to another may require extensive changes in personnel because people feel more comfortable with one kind of system than another.

Existing Relationships and Loyalties

Just as the political system reveals much about how power is wielded in an organization, the "shadow" system reveals the web of personal relationships and loyalties in an organization. The "shadow" system depends on such things as past history, personality, favors and obligations, and friendships and animosities, which have a more powerful influence on behavior than the outward political system. In specific situations, they often overwhelm larger-scale, more diffuse edicts of the political system. Organizational changes must therefore be made in light of the relationships among the individuals who play key roles.

The examples the committee examined are replete with illustrations of the role of personal relationships in governance. For example, a senator and governor from Maryland were instrumental in creating a vision for regional management of Chesapeake Bay and in motivating others to help make it a reality. The MMS regional director in southern California and the manager of the Florida Keys NMS have already been mentioned as key figures in the case studies. In these three instances, the "policy entrepreneurs," rather than feeling constrained by the limitations of available governance mechanisms were able to use them and elaborate them to achieve unusual results. They were able to do so because of their personal credibility and their networks of personal relationships. In the Gulf of Maine case study, by contrast, no dominant personalities emerged. In this case, no visible leaders arose to motivate and direct changes in existing governance systems.

Attitudes toward Innovation and Risk

Individuals, organizations, and groups of organizations (i.e., governance systems) differ markedly in their attitudes toward innovation and risk taking. Such attitudes are not permanent but are strongly influenced by prior experience, as well as by organizational missions and reward systems. For example, many corporations now attempt to encourage managerial risk taking by changing the criteria for rewarding job performance. They may, for example, put a greater percentage of a manager's pay "at risk" (i.e., dependent on bonuses and stock options) or include, and reward, risk taking as a core value. In the governance context, the attitude toward innovation and risk is important because changes involve new modes of behavior that, precisely because they are new, entail some risks. If reward systems discourage innovations and punish risk takers, changes will be harder to implement. In the southern California OCS, for example, the MMS regional director was perceived as a risk taker who was operating on the edge of the MMS culture. In contrast, innovations in the Long Island Sound case were encouraged by specifying only end goals for nitrogen reduction and leaving the methods up to local treatment plants.

Risk taking can be fostered or inhibited by a wide range of organizational characteristics. For example, national conferences and workshops for NEP managers have helped to create a support network so managers feel less isolated and have a vehicle for trading stories about successes and failures. The Maine lobster fishery hopes to adapt the overall management plan to the specifics of each local area by leaving a range of decisions to the discretion of local management councils.

Attitudes toward Information Ownership

Information is always a key currency in organizations. It costs resources to develop; it is often crucial in decision making; and it can be used to support allies and attack rivals. It is not surprising, therefore, that the way information is controlled reflects the organization's theory of ownership and control. In some cases, information is considered the property of those who create it; at the other extreme, information can be viewed as belonging to the organization as a whole or to society at large. Access to and control over information are important for moving toward more collaborative decision making. For example, repeated assurances that all stakeholders had access to the same information was an important element in building trust in the southern California OCS case study. In a slightly different way, assurances that ongoing discussions about program design would not be portrayed as "done deals" were important in reaching consensus in the Santa Monica Bay Restoration Project.

The ways information is shared within and among organizations also depend on the other axes, especially the political and reward systems and the networks of personal relationships and loyalties. There are usually important differences between an organization's overt policies toward information ownership and the ways these policies work in day-to-day practice. Understanding both policies and practices is crucial to changing organizational behavior, which inevitably involves changing the production, control, movement, and use of information.

Morale

Morale is a nebulous concept but can be loosely defined as the amount of credit, slack, energy, or resilience available to members of an organization to help them tolerate the stresses of change or uncertainty. Morale is difficult to quantify but is crucially important. In the Alaska by-catch case, morale temporarily rose as the result of an opportunity for stakeholders to cooperate on shared problems outside the constraints of the normal management arena. In the Florida Keys, a series of ship groundings raised morale in the sense that it energized residents of the Keys and motivated them to address common problems. In the Maine lobster case, fishermen's sense of attachment to specific fishing grounds increased their desire to see effective fisheries management implemented and their commitment to working through the legislative process.

The level of morale should influence decisions about how changes are promoted. Sometimes, the process of organizational change itself can raise or lower morale, often in unpredictable ways. In some cases, changes simply grind to a halt for no apparent reason when the morale "bank" runs out.

The axes discussed above do not operate independently. Because they interact in complex and subtle ways, the behavior of a governance system is "overdetermined," that is, "it is the product of multiple, nonindependent factors whose influence depends in part on the fact that they *are* redundant" (Hackman, 1990). Changes in one axis necessarily cause changes in other axes. For example, shifting from a monarchical to a federalist political system in the Santa Monica Bay Restoration Project entailed making parallel shifts in access to information, reward systems, and attitudes toward innovation. In addition, organizations are not homogeneous with respect to these axes. Different organizational parts and/or levels, as well as different individuals, express each of these axes in different ways. Thus, an organizational culture is a diverse patchwork.

Perhaps most important for improving marine area governance is an understanding of the differences between organizational structures. All of the issues and relationships come into play as the parts of a governance system interact. The southern California case study is in large part a chronicle of shifts and readjustments in power among federal, state, and local entities. One reason MMS was successful in this turbulent environment was its persistence in building personal relationships, information sharing pathways, and implicit reward systems that bridged the gaps among these various entities. Conversely, in the Oregon rocky shore case study, the state's ability to keep federal agencies at bay was crucial to providing the freedom and flexibility necessary to create a regional planning mechanism.

ALTERNATIVE MODELS OF GOVERNANCE SYSTEMS

Traditional Bureaucratic Model

The committee found that, until recently, governance and management in marine areas were usually organized and carried out according to a traditional bureaucratic model. This model (see Box 4-2) grew out of the 18th century Enlightenment view of nature as a collection of mechanistic systems with interchangeable parts that behave in understandable and predictable ways (Binswanger et al., 1990; Botkin, 1990; Sahl and Bernstein, 1995). Originally described by Max Weber (1947), the bureaucratic organization has standardized responsibilities, qualifications, communication channels, and work rules, as well as a clearly defined hierarchy of authority (Mintzberg, 1983).

Classic hierarchical bureaucracies are ill-suited to addressing the complex problems currently confronting governance and management systems. Traditional bureaucracies attempt to solve problems of coordination by increasing standardization and proliferating rules. When unusual problems arise, their only recourse is direct supervision.

There are at least three basic problems with classical bureaucracies. First, in unstable environments, where nonroutine events frequently arise (as they did in all of the examples the committee examined), the management levels of a bureaucracy become bogged down in supervising, adjudicating, and making new rules. This undermines their ability to think strategically or adaptively about the situation as a whole. Second, rigid, hierarchical pathways for the flow of information effectively prevent managers from receiving critically important information about how the environment is changing and how the organization is responding. Finally, the ability of a traditional hierarchy to deal with the "human elements" described in the preceding section is extremely limited. In short, bureaucracies are fundamentally nonadaptive (Mintzberg, 1983; Heydebrand, 1989).

With the exception of the management of fisheries in the Gulf of Maine, none of the marine governance and management organizations examined for this study functioned like a traditional bureaucratic system. Participants were virtually unanimous in stressing the need for more flexibility, greater local involvement in formulating problems and making decisions, better access to information, and novel power-sharing arrangements. They had created new governance mechanisms (Chesapeake Bay), taken risks within existing management structures

BOX 4-2
Characteristics of Traditional Bureaucracies

- highly specialized
- routine operating tasks grouped by function
- coordination through standardization of procedures and direct supervision
- formal, hierarchical chain of authority
- relatively centralized power and decision making
- strong emphasis on division of labor in all its aspects
- very formalized procedures in the operating core
- proliferation of rules and regulations
- formalized communication throughout the organization
- elaborate administrative structure
- emphasis on controlling and eliminating uncertainty
- emphasis on smooth, efficient operation
- best suited to stable, predictable environments
- ill-suited to developing realistic strategies
- fundamentally nonadaptive

Source: Mintzberg, 1983.

(southern California OCS), implemented democratic decision making (Maine lobsters), given local agency representatives greater autonomy (San Francisco Bay Demonstration Project), established collaborative decision-making processes within an existing bureaucratic permitting framework (Santa Monica Bay Restoration Project), and devised new forums outside the existing management structure (Alaska by-catch). Success in each case was attributable, in large part, to a willingness to abandon the traditional bureaucratic approach. In striking contrast, the Gulf of Maine case study was notable not only for the loss of fisheries resources but also for the absence of attempts at innovation within the existing management structure.

The committee does not mean to attack classical hierarchical bureaucracies, some of which are highly efficient and effective when the environment is stable and problems can be dealt with through standardized tasks. However, the committee's review clearly showed that most governance situations in the rapidly evolving marine environment do not meet these criteria. In an analysis of organizational failures, Miles and Snow (1992) identify two underlying causes: logical extensions of existing structures that push the structure beyond the limits of its capability, and modifications of structures, which, although reasonable on the surface, violate the basic operating logic of the structure.

Thus, many governance problems arise when bureaucratic organizational models are applied in situations for which they are fundamentally inappropriate. The remainder of this chapter discusses alternatives to classical bureaucracy that might be more appropriate for marine area governance and management problems.

Organizational Alternatives

In the past decade, the growing awareness of the limitations of bureaucracies and an expanding body of research into organizational structure and behavior has led to the development of many different organizational models. These include structured divisions, various kinds of networks, adhocracies, "virtual" organizations, federalist arrangements, dispersed nets, and ephemeral structures that come and go depending on circumstances (e.g., Mintzberg, Rochlin, 1989; 1983; Eisenberg, 1990; Roberts, 1993). Each model offers a different solution to the common problems organizations face, including:

- making decisions
- adapting to changing circumstances
- distributing power
- controlling and/or coordinating activities
- moving information and knowledge
- establishing standards
- learning from experience
- ensuring reliability
- resolving and/or suppressing conflicts
- surviving over time

Although recent research has validated a wide range of potential solutions to these problems, it has not confirmed that any one of them is the right approach. Not surprisingly, different organizational forms are best suited to different situations and to different participants. However, the situations in the examples in this study do not cover the full range of possible situations. In the examples, extreme operational reliability, rapid and sustained innovation, precise coordination of far-flung activities, and rapid, real-time crisis management were rarely needed.

The primary shortcomings of the existing marine area governance systems are related to the accumulation of laws, regulations, policies, and practices at the federal, state, and local levels and the array of agencies that try to implement and enforce them—operating in many instances with conflicting mandates. In many cases, federal policies and actions are controlled from Washington with little understanding of local conditions and needs. No mechanism exists for establishing common visions and common objectives for individual coastal sectors. These shortcomings often lead to a lack of accountability, rigidity, lack of creativity, and continued conflicts among stakeholders.

In contrast, the majority of successful governance examples the committee examined exhibited the following properties:

- a guiding, not controlling, role for centralized authority
- a high level of concern, involvement, and initiative at the local level
- delegation of problem-solving authority and power to the appropriate local level
- open access to and sharing of information
- collaborative decision-making processes

The characteristics listed above mimic the distinctive characteristics of successful management systems in other contexts where complexity and uncertainty predominate (e.g., Rochlin, 1989; Eisenberg, 1990; Hutchins, 1990; Weick and Roberts, 1993; Roberts et al., 1994). Nevertheless, it is important to recognize that these characteristics, which sound simple in the abstract, can be ambiguous and complex in real situations. It is not always clear, for example, where the centralized authority is (or should be) among the federal, state, and local stakeholders involved in complex governance problems. In addition, sometimes the same agency plays multiple roles depending on the stage of development of the governance process or the decision being made.

For example, in the larger context of the NEP in Santa Monica Bay, the Regional Water Quality Control Board is a local decision maker. However, in the specific realm of permitting, compliance, and monitoring, the board is the central, coordinating agency, and decisions about specifics are the responsibility of other local decision makers. Similarly, the state of Maine behaved as a centralizing influence when it set trap limits for lobster fishermen throughout the state but behaved as a local decision maker when it negotiated a power-sharing arrangement with NMFS giving the state wider jurisdiction over the lobster fishery. Improving marine area governance will depend in large part on creating organizations with the five characteristics described above in the context of inherently complex specific situations.

Dealing with Chaos

Another key lesson that emerged from the examples the committee examined was the chaotic and unpredictable nature of successful governance processes. A striking observation by several participants was that although, in hindsight, they could be perceived as following a carefully planned path, at the time, they were involved in confusing, often chaotic, processes that required intuition, improvisation, and even luck. The participants also stressed that governance problems were never completely solved but were dynamic and required ongoing management and adaptation. In short, establishing an appropriate organizational model does not obviate the chaos inherent in the governance and management of complex problems. In fact, they may even add to unpredictability because they dispense with predetermined scripts and leave much of the decision making to local participants.

Therefore, eliminating the chaos and unpredictability in successful examples of governance is neither possible nor desirable. These characteristics are inherent in the very nature of complex governance problems and their solutions. In fact, efforts to reduce unpredictability through tighter planning, control, and supervision (i.e., more bureaucracy) can only make matters worse because they tend to restrict the flexibility, adaptability, direct local involvement, and the flow of information.

Recent research, as well as findings from the case studies, suggest that creative problem solving in complex governance situations is greatly enhanced by the freedom to improvise within appropriate structures (Isenberg, 1985; Eisenberg, 1990; Bigelow, 1992). Research also strongly suggests that improvisation in decision making is most effective when it is based on a few generic precepts or rules of thumb rather than detailed procedures and requirements. Generic precepts provide direction, stimulus, and structure for creating solutions out of the particulars of a given situation.

The following keys to success may be drawn from the case studies and other related experience:

- Plan carefully, but be prepared to adapt and improvise.
- Foster strong leadership, but depend on collaboration and shared decision making.
- Use existing institutions, but take advantage of the opportunity to create new forums.
- Use the existing authority of federal agencies as a starting point for encouraging them to behave in new ways.
- Information should be validated by one or a few credible sources but distributed as widely as possible.
- Encourage open and direct communication, but create multiple pathways for conversation and the flow of information.

The apparent contradictions in these statements are only contradictions in terms. Classical organizational structures and functions have fixed hierarchies of elements that act in predictable ways through simple, additive chains of cause and effect. The contradictions resolve themselves when governance systems are understood to be composed of shifting hierarchies that constantly interact in ambiguous ways through complex feedback mechanisms.

For example, the sanctuary designation process used in the Florida Keys proceeded through a series of well defined steps that provided a solid structure for the planning process. Similarly, the NEP in both Long Island Sound and Santa Monica Bay laid out specific planning steps that furnished overall direction and guidance. However, these planning frameworks were only starting points that provided motivation and direction but did not determine the details of day-to-day, or tactical, activity. Local managers had plenty of room to tailor the processes to their needs. Perhaps the clearest example of adaptive planning the committee examined was by the regional MMS office in southern California, which explicitly planned for adaptability, even going so far as to train personnel to function effectively in an environment of uncertainty and collaborative decision making.

Strong leadership is a necessary prerequisite for adaptive planning. The committee observed that the most effective leaders were capable of articulating and "selling" a clear picture of improved governance and were willing to share power in order to achieve it. For example, NOAA's NOS developed clear policy guidelines that supported the San Francisco Demonstration Project's efforts to provide better services to local industry and government. Often, strong leaders were able to build on the resources and authority of existing institutions. For example, regional regulatory agencies in both the Long Island Sound and Santa Monica Bay Estuary programs used their regulatory authority to modify traditional relationships with dischargers, thereby creating a fundamentally different environment for solving problems.

Finally, successful governance typically fostered a distinctive style of communication among stakeholders. Gathering and using valid and credible information was emphasized. But distributing this information as widely as possible, through a variety of channels was emphasized. Thus, several agencies cooperated to develop a single body of scientific information for assessing the potential impacts of oil drilling in the Santa Barbara Channel in southern California. Similarly, a key aspect of the in-shore lobster management regime in Maine was the availability of scientific information useful to fishermen. Along with wide distribution of information, there was a willingness to create a variety of formal and informal (and often transient) forums for using it. This willingness created a climate for many kinds of useful discussion (e.g., exploration, negotiation, problem solving), often at the same time. Under these circumstances, an organization can anticipate and deflect crises rather than simply react to them (Rochlin, 1989).

Sincere, sustained efforts are already under way within federal and state agencies to explore new organizational models and new approaches to governance. However, their efforts have been hampered in many instances by prevailing mindsets based on the assumptions of classical hierarchical bureaucracies. Although the successful governance structures in this study did not share a single organizational model, they were all characterized by a redefinition of roles, particularly of federal agencies, and a redistribution of power. Action was initiated and/or directed by people from the local area who understood the local problems and were determined to improve conditions by working together across existing organizational boundaries. A central guiding force was present in most examples the committee examined, but the bulk of the detailed, day-to-day decision making in the successful examples occurred at the local level. In other words, governance in these cases could be characterized as federalist, the same governance mode that characterizes the overall national structure.

5

Improving Marine Governance

In the last decade, a new paradigm has emerged for governing natural re-sources and the global environment. This paradigm was first set out in a seminal report by the United Nations Commission on Environment and Development—the Brundtland Commission (1987). The report presents a powerful argument that the goal of environmental management in the future must be integrated into an overall strategy for sustainable development (or sustainable use). The Brundt-land report points out the inextricable links between the problems of environmen-tal degradation, poverty, overpopulation, and unsustainable development, all of which must be attacked together. The goal of sustainable development is further defined by the ethical principle that actions taken today must not compromise the environmental and development options of future generations.

Although the Brundtland Commission's work was cast in a global context, the principal recommendation concerning the need for sustainable development applies to the United States and has been further developed in a report by the President's Council on Sustainable Development (President's Council on Sus-tainable Development, 1996). The collapse of the New England ground fishery is just one example of poor governance. Another example is the population growth on many coastal barrier islands which, over the long term, will have deleterious effects on fragile, near-shore marine plant and animal species and will make the islands themselves more vulnerable to climatic events and changes.

Some unique characteristics of marine areas make managing them extremely difficult:

- Underwater lands and marine resources are generally publicly owned; conflicts about property rights do not apply; all of the problems of managing common resources apply to marine areas.
- Marine areas more than three miles offshore are under direct federal jurisdiction; federal involvement in their management is legally required.

A key condition for the sustainable development (or use) of natural resources is an integrated approach to management that takes many factors into account. Unsustainable development often occurs when the governance framework is incomplete or limited—that is, when not all of the interest groups are adequately represented. The unsustainable use of marine resources is sometimes also related to a lack of good scientific information or understanding or, in some cases, a tendency by those responsible for management to yield to political forces.

Better integrated governance is essential for the coastal and marine areas of the United States. Many problems in the present ocean governance arrangements (e.g., the virtual stalemate in the offshore oil and gas program) can be traced, in part at least, to fragmentation in the present system. Each ocean program operates under its own legislation and its own regulations, which are administered by an office or agency dedicated to that particular activity. A few durable mechanisms have been put in place, however, for coordinating policy, identifying and resolving conflicts early, and making rational trade-offs. These include:

- the Coastal Zone Management program in 30 coastal states and territories, under which focal points have been created to harmonize state, local, and federal activities with state coastal policies and good coastal practices
- the National Marine Sanctuary program, under which area-wide marine management programs are put in place in important ocean areas off the U.S. coast
- the National Estuary Program, under which multi-agency and multi-jurisdictional approaches are being developed on a water-body-wide basis for major estuaries in the United States

These three programs demonstrate that it is possible, under the existing legislative framework and in certain situations, to improve marine area governance.

For reasons discussed throughout this report, the most appropriate focal point for improving marine area governance for the United States appears to be the regional level (e.g., the New England region or the Gulf of Mexico region). However, broader national interests in the ocean also need to be articulated and protected, and this can best be done at the federal level (see discussion in Chapter 2).

National interests are spelled out, at least partially, in legislation and/or in the Constitution, but no authoritative guidelines have been established for setting priorities or policies in the case of conflicts between and among these interests. Nor is there a mechanism for identifying when national objectives are not being met. The committee believes a new mechanism is needed at the federal level to consider and address these problems. The same mechanism could move beyond these substantive national interests to address some fundamental problems in the current governance system.

Effective governance requires mechanisms for allocating authority among federal, state, and local governments and continuous, sustained coordination among federal agencies. Large-scale problems (in terms of both area and time) and appropriate responses to them must be effectively identified before ecological crises or irrevocable damage has been done. Information and data need to be managed better and made available to the public so there can be constructive public involvement in decision making and so that all participants can be held accountable for their actions. Finally, systems need to be developed at the federal level to enable (and encourage) individuals and agencies to take calculated risks to improve governance.

FEDERALIST MODEL

The committee's analysis of the case studies and other coastal programs shows that governance and management often suffer from fragmentation, a lack of accountability, rigidity, a lack of creativity, and conflicts among stakeholders (see Chapter 3). At first glance, solutions to these problems appear to work at cross purposes. For example, coordination (the cure for fragmentation) can require centralization; creativity and accountability may require decentralization and the delegation of decision making authority. Neither solution will work in isolation. A superagency is likely to become bogged down in its own bureaucratic processes and is likely to be insensitive to important regional variations. At the other extreme, leaving all decisions up to local entities would make it difficult or impossible to identify and respond to larger national interests, to establish common standards, or to manage resources that extend beyond the boundaries of a particular region.

Reconciling these apparent contradictions requires a hybrid system with a centralized structure to coordinate the activities of various agencies and serve the national interest. The same organization must have decentralized mechanisms for making decisions based on the local understanding of problems. This unique system could be modeled on the federalist system, which is well developed in the management of land-based resources.

In selecting the federalist model as the defining concept for reinvigorating marine governance, the committee is not embracing the historical use of that term, which is understood as the advocacy of a strong central government at the expense of the states. Rather, the committee is using the term in a more modern sense to mean a governance system based on true sharing of authority and responsibility between the national government and the states. This modern federalist model would give the states critical authority for carrying out many of the functions of government for which they are ideally suited because of their proximity to the problem areas (Hershman et al., 1988; John, 1994). The committee also suggests that a federalist governance structure could be useful in areas beyond the three mile limit where there is no local partner to share governance responsibilities with federal agencies.

A federalist marine governance system would have the following basic components:

- a National Marine Council to coordinate the activities of federal agencies, define national responses to global problems, monitor the marine environment, facilitate regional solutions to marine problems, and encourage interagency problem solving
- regional marine councils to address complex governance issues at the operational level
- conditions that make it easier for individual federal programs to succeed
- a wide range of management tools that would make regional councils and individual agencies more effective

National Marine Council

The role of the central governing body in a federalist model differs from the same role in a hierarchical, command-and-control model. In the traditional model, the center decides, controls, mandates, enforces, and punishes. In the federalist model, the center sets boundary conditions, organizes, persuades, motivates, and ensures that fundamental issues are raised and resolved. In the federalist model, the center holds participants accountable through motivation rather than by punishment.

Previous efforts to improve the effectiveness of government fall into two broad categories—procedural and structural. Procedural reforms have included attempts to enact integrated statutes covering the marine realm and extending the scope of environmental impact assessments required under NEPA. Drafting integrated statutes under NEPA has proven to be a daunting task. The environmental impact assessments provided for in this legislation cannot be easily extended beyond their present scope. This statute applies to federal activities in the marine environment but does not provide a coherent framework for governance or management (John, 1994).

Structural approaches to change include a range of alternatives. At one extreme, existing agencies and authorities at all levels of government could simply be exhorted to do their jobs in new and better ways. Although this strategy is appealing in its simplicity, it has rarely worked in settings as complex as natural resource management where many public and private interests are involved. Exhortation, unaccompanied by organizational changes, rarely leads to implementation and is likely to fade away with changes in political administrations or philosophies.

At the other extreme is the option of creating a superagency that integrates the functions of various federal agencies and provides for overall governance of the ocean. However, creating a new organization within the federal bureaucracy is hardly feasible in the present political climate (National Performance Review, 1993). Also, reorganization is generally very costly and seldom yields significant benefits (e.g., the Energy Research and Development Administration, the Department of Energy). In the case of marine governance where the paramount objective is more involvement by state and local interests, a superagency might even be counterproductive (Cicin-Sain et al., 1990).

The committee recommends a third option for consideration, the creation of an interagency council—a National Marine Council—composed of leaders of key ocean agencies. The National Marine Council should be a permanent body created at the direction of the White House and should report directly to the President. The council's functions are suggested in the following discussion. Council members should be senior policy representatives of the key federal agencies responsible for marine affairs. At a minimum, these would include the U.S. Department of Commerce (NOAA), the U.S. Department of the Interior (U.S. Geological Survey, MMS, National Park Service), the EPA, the U.S. Department of Defense, and the U.S. Department of Transportation (the U.S. Coast Guard and the Maritime Administration). The National Marine Council should have a permanent staff with a modest budget and a chairman appointed by the President to carry out its mandates.

There is a danger that the National Marine Council could create another level of bureaucracy in an already crowded arena, but the committee believes that the tasks defined for the council will effectively improve governance and management and that the cost in resources and increased government presence will be outweighed by the benefits.

The committee envisions the National Marine Council as a body with decision making authority and accountability for the state of the nation's oceans. The council should not merely be a coordinating mechanism, although one of its major functions would be to coordinate and harmonize the activities of various federal agencies. The council's responsibilities obviously could not exceed the authority of the individual agencies or the President's overall policy direction, but within the authority of existing laws, there is a great deal of scope for agencies to improve governance of the marine sphere, as described in this report.

Two tasks (outlined in detail below) would be critical to the council's success: (1) establishing national priorities and reporting on the status of the marine environment; and (2) engaging a wide range of local interests and skills to solve problems. This latter function would be carried out in conjunction with either existing entities involved in marine activities or with newly created regional marine councils (described below). In either case, the National Marine Council would empower the regional structures to define and solve problems in innovative ways. The objective would be to cut through the rigid, slow processes of centralized authority and maximize the effectiveness of local actions to achieve national objectives.

A National Marine Council could develop goals, principles, and policies for resolving problems as they develop. The council could also review existing federal legislation with an eye toward developing simple, understandable definitions of the national interest as it pertains to water quality; biological diversity and abundance; national security; human health and safety (such as navigation safety, protection from coastal and marine hazards, and seafood safety); and other legitimate national concerns, including economic development.

In compiling this information, the National Marine Council could express the national interests, insofar as possible, in measurable terms to facilitate assessments of the success or failure of governance arrangements. Furthermore, these goals and policies would not add a level of regulation but would consolidate, simplify, and integrate goals and the standards already set out in legislation. New national goals, however, may need to be developed in response to unanticipated situations or to establish an authoritative process for setting priorities among national goals or interests. The National Marine Council would also have a coordinating and decision-making role in implementing the overall marine management strategy.

Past experiments with federal marine councils focused on science, technology, education, and interagency coordination, but they were not directly involved in management and governance. The National Marine Council would have a much broader mandate. Its job would go beyond coordination to include setting national priorities; developing a multi-agency plan for monitoring and reporting on the environment; identifying potential large scale problems or issues; and holding federal agencies accountable for failures of governance. Most important, the National Marine Council would encourage attempts to implement regional approaches to marine management.

Role of the National Marine Council

As an effective partner in a strengthened federalist system, the National Marine Council should perform the following functions:

Define the National Interest in Marine Areas. Defining the national interest in marine areas is central to changing the current marine management system. Although numerous constitutional, executive, and congressional pronouncements have been made regarding the federal role in managing the marine environment, they do not add up to a national agenda with priorities for scarce federal resources. The National Marine Council should define the nation's strategic priorities for sustainable use of the marine environment on a regular basis. The National Marine Council should take into account the risks and benefits of marine resource use and development and prioritize allocation strategies among the competing interests.

The council's primary objective should be to ensure that the United States has clearly identified the global and national issues that require the mobilization of the resources of the federal government. Some criteria have already been developed, such as the water quality standards established under the Clean Water Act and national criteria for managing striped bass. National criteria are, in effect, performance standards that protect the national interest in marine areas. The council should develop criteria for determining when federal agencies should assist regional and local managers and when federal agencies should take direct action to protect resources. National criteria and standards should still be established by existing agencies under existing authority, but the National Marine Council should coordinate activities among agencies to protect national marine priorities.

Monitor and Report on the Marine Environment. Federal, state, regional, and local agencies have developed a variety of programs and projects to monitor the condition of the marine environment. These programs often overlap or leave gaps in monitoring. The National Marine Council should encourage and facilitate monitoring of both physical conditions in coastal waters and human uses and conflicts, on a regional basis.[1] The council could do this either by coordinating communication among federal agencies or by delegating monitoring responsibilities to regional agencies. Data from fisheries and biological data from other sources—including from the academic research community—should be included in a nationwide monitoring network.

The National Marine Council should compile and publish monitoring information that highlights environmental progress or indicates risks to marine resources. In other words, this document should be a report card of coastal conditions.

Identify Marine Area Problems and Conflicts. Monitoring and communications systems should be put in place to identify problems as early as possible. Once problems were detected, appropriate action should be taken by federal or state agencies before irrevocable damage has been done. If the problem threatens to interfere with achieving national goals or with the mandate of a federal agency, the council should make sure that the appropriate agency takes action. If the problem is within the mandate of an existing federal agency, the council should work directly with that agency. If the problem is regional but contradicts national priorities, the council should work with local authorities to develop an effective response or to determine whether a regional council should be established.

Protect the National Interest If Regional Efforts Have Not Addressed Significant Risks to Important Resources. The council should intervene in a situation only to address an unacceptable risk to an important resource and do the minimum needed to resolve the problem. Stakeholders must always be involved in seeking solutions to problems. The council should also intervene if a single-purpose federal agency fails to protect resources and benefits of marine waters.

---

[1]See NRC, 1990b for an examination of the role of marine environmental monitoring.

Intervention by the council could simply mean working with a federal agency to implement existing statutory authority.

Encourage Regional Innovation. A major purpose of the National Marine Council would be to provide federal support for regional management. Experiments in regional governance can only be tried if the federal government is willing to make them work. The National Marine Council should be a vehicle for the cooperation and involvement of relevant federal agencies. The council should create a climate within the federal government that allows for innovation and risk taking by federal authorities at the local level.

Regional Responses to Marine Area Governance

The case studies show that the existing governance system is complex and often involves regional offices of federal agencies that have varying degrees of autonomy from their Washington headquarters. Each federal agency has a different approach to management. For example, the NMFS works closely with regional fisheries management councils in establishing policies for fisheries conservation. In the case of the MMS, some regional offices are tightly bound to headquarters policy and directives, but others are not. In the case of the Florida Keys NMS, NOAA headquarters was heavily involved in the creation of a management plan. Each of these approaches involves federal agencies interacting with each other, state governments, and local interest groups.

This complex governance system has achieved some notable successes, but in too many instances, it has failed to anticipate problems or to come up with timely solutions. The failures fall into two basic categories. Either federal and state or regional programs and interest groups have attempted to solve a management problem, but existing institutions and practices have impeded them, or federal and state or regional organizations and interest groups have not recognized and reacted to marine management problems, thus jeopardizing national and local interests in a marine area.

In many areas of the country where marine resources are subject to risk or conflicts, citizens, stakeholders, local and state governments, and the regional offices of federal agencies may recognize problems and address them effectively. If existing programs can address problems and risks, they should be used. But if federal regulations and missions conflict, if federal and state or local interests conflict, if there are not enough resources or expertise to address the problem, or if the situation is unusually complex, a new process may be necessary.

Role of Regional Marine Councils

In situations where there are complex or long-standing conflicts or risks, the committee recommends the creation of regional marine councils representing federal and state agencies and other stakeholders. Members should be appointed by the National Marine Council through nominations by state governors and regional federal agency heads (similar to the way estuaries are designated under the NEP), or through interagency discussions within the National Marine Council. This second method would be used if federal agencies disagreed about the effectiveness of the management of a marine area or resource. The composition of each council would depend on the problem and the region. Regional councils should be created only in areas where there are substantial risks of damage to highly valued resources or serious, long-standing conflicts. Regional councils could be established at any time and should remain in existence for the duration of the problem but should not be permanent. Regional councils should be established through formal agreement between the council members and the National Marine Council.

At first glance, the regional fishery management councils created under the FMCA (Fishery Management and Conservation Act) might seem to be analogous to regional marine councils. The new regional marine councils, however, would be structured to overcome two problems that limit the effectiveness of the fishery councils. First, the mandate of the fishery management councils under the FMCA is limited to maximizing the exploitation of the fisheries resource. They often fulfill their mandate to the exclusion of a variety of other socially desirable purposes, including the long-term survival of commercially valuable fisheries stocks. Second, the fishery management councils did not include representatives of a broad range of interests concerned with the overall health and utilization of the marine environment. The regional marine councils proposed in this report would have broader objectives and goals and would represent a broad range of interests.

The regional councils have some elements in common with the regulatory and management structure described in a recent analysis of the changing nature of environmental policy making (John, 1994). This study suggests that the federal role be changed from commanding, controlling, and enforcing to setting priorities, providing information, and furnishing incentives. Neither that study nor this committee advocates abandoning the federal role of setting and enforcing standards. But they agree that more effective governance depends on problem-solving based on local knowledge and local participation. At the same time, the committee recognizes that the dynamics of a successful regional council are extremely difficult to predict. For this reason, existing institutions that have been successful in the past should be tried first. Regional councils should be created only where the nature of the problem and the interaction of the interests clearly require a new structure.

Areas where regional councils should be considered should have at least some of the following characteristics:

- geological or biological features that define them as regions in scientific and political terms. (In the case studies, the Gulf of Maine and Florida Bay and the Florida Keys are logical geographic areas for regional councils. A breakthrough was made in planning the restoration of Chesapeake

Bay when elements of the entire bay watershed became part of the planning unit.)

- problems and conflicts that cannot be readily resolved by conventional means (e.g., the depletion of groundfish stocks in the Gulf of Maine)
- resources that are particularly important to the future of the region and the nation that may be jeopardized without timely planning and management. (In southern California, scenic/ecological and mineral resources of national importance could be lost without effective planning and management).

In general, regional councils should encompass whole coastal ecosystems regardless of political boundaries. They should be chaired by a lead agency and should have members representing state environmental agencies, federal agencies with direct responsibilities in the area, and key stakeholder groups. Participants must have decision-making authority or ready access to those who do. In some instances, this means that senior agency staff should be directly involved. In others, it could mean that elected officials or their senior aides should participate. It may be necessary for stakeholder groups to elect representatives who have the authority to present their case and commit to solutions. Depending upon the complexity of the problems, a regional council might designate subcouncils to deal with specific aspects of the overall problem or to provide technical advice. Once established, regional councils could perform the following functions:

Develop Long Range Goals and Plans. A council could be a forum for identifying problems, setting priorities, establishing long-range goals and developing plans to attain those goals.

Coordinate Planning and Management among State and Federal Agencies. A council would coordinate agency responses. The South Florida Ecosystem Working Group, which is establishing priorities for restoration of the Everglades, is a terrestrial example. The existing Gulf of Maine Council is another example, but it has less real authority than the Ecosystem Working Group.

Coordinate Fiscal Planning Including Pooling Funding from Two or More Government Programs or Agencies. Management of funds is often an obstacle to effective interagency coordination and to using available funds in the most cost-effective way. Pooled funding should include simplified performance and reporting requirements. (There was some coordinated fiscal management between EPA and NOAA in south Florida, but more could have been done to integrate water quality and habitat programs).

Mediate and Resolve Disputes among Agencies and Stakeholders through Environmental Mediation and Related Tools. So-called alternative means of conflict resolution have been used successfully to resolve environmental disputes or impasses to environmental action. Regional councils should sponsor alternative means of conflict resolution. Federal agencies could pool funding to hire mediators and facilitators. (The Florida Keys NMS might have benefitted much earlier from mediation among conflicting stakeholders. In southern California, environmental mediation techniques were successful. The Alaska Fisheries By-Catch workshops were attempts to use alternate dispute resolution techniques to bring stakeholders together.)

Facilitate Intergovernmental Agreements. Councils should facilitate formal intergovernmental agreements, some of which may result from environmental mediation. These agreements would bind the signatories to managing marine resources in a certain way. (Intergovernmental agreements were important in Chesapeake Bay and Long Island Sound for formalizing interstate cooperation on pollution reduction.)

Waive Some Regulatory Requirements to Achieve Performance-Based Goals. Overlapping and conflicting regulatory requirements are sometimes obstacles to resolving environmental disputes and to taking proactive steps to avoid damage to resources. Regional councils should have the authority to negotiate memorandums of agreement that waive some regulatory provisions in order to achieve performance goals. Memoranda should be subject to approval by the National Marine Council to ensure that they do not violate the spirit of underlying statutes.

Execute Stakeholder Contracts. In the course of the study, the committee learned that marine interests and stakeholders must be responsible for, and a part of, the resolution of user conflicts and efforts to repair damage to marine resources. One mechanism for involving users is through contracts that establish shared responsibility for desirable outcomes.

Performance goals should be negotiated between user groups and agencies. The goals would have to be sustainable, achievable, and in compliance with the statutes governing marine areas. Once goals have been established, a contract would assign the management of a resource to a user group (or to a user group and other interests, including state or local government) establishing benchmarks for success or failure and setting a timetable for performance. This arrangement would allow substantial freedom for the responsible party to design management measures that meet bottom-line goals. Failure to achieve the benchmarks would result in penalties or termination of the contract.

An example of contracting would be to allow a group of lobstermen (or lobstermen in cooperation with government agencies) to manage lobsters along a section of the Maine coast to ensure agreed upon survival rates and population levels. The contractors would be allowed to establish management rules (like mesh sizes, seasons, by-catch provisions) provided that the performance measures were achieved.

Engage Local Interests. Regional marine councils would be focal points for bringing together diverse groups interested in resolving conflicts in a particular region. The council would encourage responsibility and accountability, as well as create a forum for participatory democracy in the management process. Participants would have the opportunity to participate in the decision-making process and, at the same time, to assume some of the responsibility for success or failure.

Monitor and Evaluate the Results of Contracts and Other Management Actions. In conjunction with the monitoring systems developed by the National Marine Council, regional councils could coordinate and integrate monitoring by individual agencies to ensure that information was gathered effectively and cost-effectively. Monitoring would be essential to the performance-based management tools described above. (The San Francisco Bay Demonstration Project is an example of several agencies combining their efforts to monitor a resource.)

Provide Technical Assistance and Training. Participatory decision making, such as the processes proposed for the regional councils or for single agencies, often requires training and technical assistance. Regional councils could use pooled funds to train participants and stakeholder groups in effective decision-making and in interpreting and using scientific data. (In all of the case studies, participants in decision-making processes would have benefitted from training, which could have reduced conflict.)

Ensure Accountability. Sometimes the roles and interactions of public agencies in addressing environmental problems are so complex that it is difficult to hold anyone responsible for failure or reward anyone for success. If a regional council assumes responsibility for solving management problems, the council members would be accountable for the results. The National Marine Council should have the authority to dissolve regional councils if they fail to perform well and to assign remedial tasks to member federal agencies.

Advantages of a National and Regional Council Approach

The federalist model for marine area governance proposed by the committee has several benefits. It creates an organization that can coordinate federal activities and protect national interests in marine areas and provides a framework for a national, cost-effective marine monitoring system in partnership with regional and local interests. A monitoring system could identify risks to resources before crises or irrevocable losses had been incurred. The proposed council would not interfere with existing marine management programs that are working effectively to avoid or solve problems. It would encourage coalitions of federal and state agencies and interest groups to identify problems and work together to solve problems, and it would give them the flexibility to design and modify programs and processes to meet regional needs. For example, councils would be able to use contracts to enable stakeholders to govern their own actions and still protect resources of national importance. This approach also creates checks and balances among federal agencies to encourage changing a course of action when resources of national importance are threatened. A major benefit is that the council approach is based on creating coherent, cooperative relationships among existing laws and institutions rather than creating new structures.

IMPROVING EXISTING PROGRAMS

This section sets forth a variety of approaches to improving management and governance by existing programs. These approaches could be implemented either in conjunction with the two-tiered council structure proposed in this report or within the existing management and governance framework. This discussion is not intended to be exhaustive but to illustrate how the framework and processes described in this report can be used to improve existing programs and processes.

Coastal Zone Management Program

The integrity of the coastal ecosystem requires consistent management, from the upper reaches of the watersheds through the immediate coastal zone and out into the deeper waters of the coastal ocean. Although the existing CZM (coastal zone management) program is responsible for managing the principal coastal ecosystem, in some cases, adjacent marine areas would benefit from being governed in a manner consistent with existing CZM programs. The federal-state partnership principles built into the CZM program could be applied to many other aspects of marine area governance.

The proposed National Marine Council could strengthen the CZM program by providing a single, high-level location for developing national ocean policy. A national council would be a much needed focal point for coordinating federal ocean and coastal programs. The National Marine Council would strengthen the federal coastal management program by establishing an authoritative mechanism for defining the national interest, especially in situations where existing legislative mandates conflict with each other. It would also create a mechanism for coordinating federal agency programs to support and enhance state CZM policies and programs, as well as a mechanism for integrating coastal programs, such as the CZM and NEP programs, with marine programs, such as the NMS program.

Regional marine councils would be in a position to greatly facilitate coordination among state CZM programs in a given region. Regional councils would provide a forum for discussing and developing regional policies that could be incorporated into state CZM programs, as appropriate. Regional councils could also direct technical assistance to state CZM programs, as needed.

National Marine Sanctuary Program

The NMS program has several qualities that make it a strong candidate for strengthening marine area governance. First, it has been especially effective in cases where highly valued natural resources were at risk over substantial areas (e.g., the Florida Keys and Monterey Bay sanctuaries). Second, it can effectively manage a range of activities, including fishing. The NMS program has the capacity to utilize zoning as a means of managing conflicts among competing uses—a concept that this committee strongly endorses. In order to maximize the benefits of these qualities, management of the planning and governance process should be fully decentralized to the particular sanctuary.

The NMS program also needs to incorporate principles of ecosystem management that mandate comprehensive, effective management of the fish within sanctuary boundaries. Although it is not essential for the NMS to manage fisheries directly, it is essential that sanctuaries assume responsibility for the health of all resources within their boundaries and recognize that overfishing is a potential threat to living marine resources.

National Estuary Program

Although it is difficult to generalize about the success of the NEP, several aspects of the program are noteworthy. First, most NEPs do not deal with fisheries management, which is often a major aspect of an integrated ecosystem governance program. This is not surprising because the authorizing language for the NEP is the Clean Water Act. However, problems with fisheries may reflect larger problems of general decline in biological resources due to habitat degradation. Addressing problems of water quality and habitat without addressing the related problems of fisheries can be unproductive and unsuccessful.

Another limitation on the effectiveness of NEPs is that implementing a CCMP (comprehensive conservation and management plan) is purely voluntary. Because the NEP provides no significant funding for implementation, they have little recourse if goals are not met. For the NEP to be more effective at marine governance and to actively engage local interests, they will have to overcome the limitations of volunteerism.

Recent reviews of the program (Imperial and Hennessey, 1996; Martin et al., 1996) concluded that the most important change for NEP would be to fund the implementation of CCMPs. However, it is important to distinguish between two different kinds of implementation funds. Some funds ensure the continuation of the CCMP process into the implementation phase; this funding is used for monitoring, assessment, reporting, and the other activities for maintaining the institutional framework of the CCMP. This kind of funding is essential for NEP programs to move beyond the planning phase.

The second kind of funding is for direct improvements to the health of a particular estuary, including, for example, installing advanced wastewater treatment systems to reduce nutrient loadings; restoring degraded wetlands; retrofitting systems for urban storm water; and cost sharing of agricultural "best management" practices. These activities are expensive, and, unless a local NEP is actively involved in implementation, they may simply produce expensive planning documents, or, at best, palliative programs that slow, but do not stop or reverse, the decline of an estuary. A discussion of options for funding can be found in Chapter 6 (Financing Marine Area Governance and Management Programs) and in Appendix D.

The second improvement for the NEP is to institutionalize responsibilities for implementation (Imperial and Hennessey, 1996; Martin et al., 1996). The NEP does not have a planning component, and it is unclear how changes or modifications to CCMPs could be made. Martin et al. (1996) recommend that the implementation of CCMPs should be made a "non-discretionary duty of the Environmental Protection Agency," which would clearly give EPA ultimate responsibility. In contrast, EPA's 1996 draft report to Congress refers to the EPA as "just one of the stakeholders." Others have suggested that improving ecosystem management would require shifting the responsibility for environmental protection away from the federal government and to state and local governments. In this view, EPA's role would shift from setting standards to providing technical assistance (Imperial and Hennessey, 1996).

Outer Continental Shelf Oil and Gas Leasing Program

A new paradigm is needed for managing oil and gas leasing in the OCS because the present program is both ineffective and inefficient. Congressional moratoria are not long-term solutions for managing the oil and gas activities in the OCS. The Southern California case study shows that a more open, inclusive approach by the MMS led to dramatic improvements. The local task force used by the MMS in that situation was, in effect, a regional marine council as described in this report. The task force included all stakeholders—federal, state, and local—information was freely provided, trust was established, and the parties worked together to find solutions rather than rigidly defending their positions or philosophies.

This paradigm should be adopted in other areas where there are serious conflicts or disagreements over OCS leasing. In those cases, the National Marine Council should establish regional (local) marine councils, which include representatives of all interested and affected parties who have legitimate interests in the area. The regional councils should be charged with developing cooperative approaches to managing the OCS activities, based on good science and a balance among various resources and potential uses of the ocean and coastal region.

Fisheries Management Program

Institutional problems have contributed to the failure of the management of certain fisheries. An NRC report (1994a), for example, found that the lines of authority and responsibility between the Secretary of Commerce and the regional fishery councils were confusing, which created inefficiencies and generated conflicts without providing a satisfactory mechanism for resolving them. The system also lacked independent checks and balances.

The proposed National Marine Council could perform many of the functions recommended in that report. It could function as an independent oversight body that could advise the Secretary of Commerce, the fishery management councils, and Congress and provide an independent mechanism for reviewing strategic planning for fisheries, reviewing controversial management decisions, resolving interagency conflicts, and coordinating federal policies that affect fisheries investment and infrastructure. The National Marine Council could review and report to Congress on performance and problems in U.S. marine fisheries, make recommendations on certain scientific and technical issues, define management goals and strategies, and highlight emerging jurisdictional problems and environmental and conservation concerns.

The role of the proposed regional marine councils is also suggested in the recommendations of the NRC report (NRC, 1994a). At the request of the Secretary of Commerce, a regional fishery council, the National Marine Council, or an ad hoc regional council could be convened to provide a forum for conflict resolution. The council would render a nonbinding decision to resolve the conflict.

The regional marine councils could benefit the existing fisheries management system. A regional council could be a vehicle for bringing various perspectives and interests into one arena. If necessary measures (such as protecting an important habitat) can only be taken through other programs (such as CZM or the Clean Water Act), the regional council could also bring these parties into the common setting. Even without regional councils, the existing system could be improved by the judicious use of contracting (see Chapter 6) to encourage self-governance by fishermen and their communities. Broadening the membership of fisheries management councils to include more stakeholders and create a less polarized environment would also be beneficial.

6

Improving Marine Management

MANAGEMENT TOOLS

Institutions responsible for designing and implementing policies can use a variety of management tools, including command and control or direct regulation (e.g., emission limitations under the Clean Water Act); moral suasion (e.g., marine debris programs); liability and compensation (e.g., recovery of damages under the Comprehensive Environmental Response, Compensation, and Liability Act (CERCLA) and the OPA 90 (Oil Pollution Act of 1990); direct production of environmental quality (e.g., sewage treatment facilities, fish hatcheries); education; economic incentives (e.g., taxes, tradable permits, and subsidies); and tools that affect the underlying dynamics of the marine system. Many of these management tools have already been used successfully in the marine environment.

No one management tool is appropriate under all circumstances. Choosing a management tool involves weighing the historical, technical, and economic factors, as well as the social and political context of resource use. Some of the available management approaches could be used often and vigorously to prevent further deterioration or depletion of marine resources. These tools could greatly improve marine governance within the existing institutional arrangements.

Managing Conflicting Uses

Users of the marine and coastal environment are imposing increasingly heavy costs on each other and on the marine environment and services. Jet skiers, for example, create safety risks for swimmers and noise pollution that interferes with other activities; speedboats are dangerous to scuba divers and may have a negative effect on marine life, such as manatees and other marine mammals; using coastal waters for sewage and sludge disposal is inimical to recreation and compromises the quality of underwater habitats; aquaculture can interfere with fisheries and navigation and can be aesthetically unappealing; naval target practice can disturb wildlife protection areas. Economists call these costs *externalities* because they are not reflected in market transactions or cost accounting. Because organizations and individuals do not have strong economic incentives for considering externalities in their decisions, they must often be dealt with through regulation or informal sanctions. Several regulatory approaches can be used to increase the incentives for limiting environmental impacts.

Zoning/Refugia

In situations where the combined use of a resource is less valuable than single use, separating them in space or time can be useful. Zoning is one method often used to reduce externalities on land (Kelly, 1988). Especially in the near-shore environment, zoning is a relatively low-cost, effective management option for dealing with conflicting uses. In the marine and coastal environment, zoning has been used to segregate commercial, recreational, and aquacultural activities; to protect wildlife sanctuaries and the marine environment generally; and to isolate waste disposal sites. Sensitive near-shore areas are often zoned as low speed or no wake areas. Certain vessels, such as oil tankers or other carriers of hazardous cargo, may be required to use specific routes to separate them from protected features of the marine environment. Zoning is used to isolate military areas, such as bombing ranges, submarine surfacing areas, and areas that affect national security. Energy installations, such as oil production facilities, are often subject to zoning restrictions similar to waste disposal sites. Another example of zoning is Hawaii's restriction confining high-speed boating and other high-speed water sports to designated ocean recreation areas.

Marine and coastal protected areas (MCPAs) are a legislative tool for protecting marine resources in a defined geographic marine or coastal area. The primary objectives of MCPAs are to preserve marine biodiversity, to maintain the productivity of marine ecosystems, and to contribute to the economic and social welfare (Kelleher et al., 1995). MCPAs have been designated in response to emergencies (e.g., extinction of a species) or, in one case, in conjunction with a land-based park.

Liability

Making parties legally liable for the economic damages they inflict on others is another well established method for dealing with conflicting uses. Private parties who have property rights in the marine environment can sue to recover damages. For example, the owner of aquaculture net pens can sue a boater who causes damage to the aquaculture facility.

The government, as the steward of marine resources held in public trust, can recover the value of damages to natural resources from parties responsible for chemical and oil spills through CERCLA and OPA 90. Although laws are limited by the vagaries of judicial decisions and the difficulties inherent in determining fault under conditions at sea, laws establishing liability provide incentives for marine resource users to avoid inflicting damages on other users and to internalize external costs.

Compensation

Another incentive-based management tool is to create a framework by which injured parties can be compensated for economic damage. For example, offshore oil developers contribute to a fund to compensate fishermen who lose gear as a result of offshore oil and gas development. Shippers have created a fund to pay cleanup costs for accidental spills. The costs for these programs are lower than they are in the judicial process, but the incentives for avoiding damages are also weaker because the average losses of the group (e.g., shippers or oil companies) determine the amount each party must contribute to the fund. Compensation mechanisms can sometimes blunt opposition to a new development. On land, for example, developers of potentially noxious facilities have developed contingent arrangements with neighboring landowners to compensate them for lower property values or other damages.

Some funds are both international and mandated in the United States under OPA 90 to compensate for damages from accidental discharges of oil into marine waters. The International Convention on Civil Liability for Oil Pollution Damage, which has been in effect since 1975, makes shipowners strictly liable for damage from oil pollution that can be traced to their ships. Shipowners thus carry liability insurance, which is made available through a system of clubs. Further compensation for oil pollution is available through the International Oil Pollution Compensation Fund, which is funded by a tax on oil companies for their oil imports.

Prenegotiated Mitigation

In some situations, prospective users of a marine resource must complete a permitting process that allows agencies representing other interests to "sign off" on the proposed use. Examples in the marine environment include permitting for marine aquaculture facilities and for dredging marine waterways. If permits are a precondition for use, this mechanism gives other interests the power to compel the mitigation of potential damage. However, unless potential damages are combined with mechanisms for compensation, existing interests can stalemate potential new users, even if they propose a worthwhile use of the resource.

There are many marine area examples of permitting requirements for certain allowable activities. In Willapa Bay and Grays Harbor in Washington state, oyster growers are required to obtain permits to treat aquatic oyster beds with coarbaryl, a pesticide used to control populations of ghost and mud shrimp. Under the authority of the Clean Water Act and the Rivers and Harbors Act, permits are needed from either the U.S. Army Corp Engineers or the EPA (often both) for changes to wetlands that entail dredging or filling, as well as for shoreline-hardening construction, such as bulkheads, groins, docks, and walls. The discharge of wastewater into the marine environment requires permits from EPA depending on the treatment methods and the condition and effects of the discharge on the receiving waters.

Controlling Access to Marine Resources

Many management tools are available for controlling access to marine resources by strengthening the rights of defined groups or individuals. One approach to the problem of common property is to establish exclusive, enforceable private property rights to the resource. Fish provide the best example. A private owner is granted the right to receive the full benefits of conservation and enhancement, to exclude others from taking those benefits, and to sell or lease his or her rights voluntarily.

Private use rights may be feasible for in-shore shellfish fisheries, for anadromous species fisheries, and for aquaculture fisheries. Indeed, many jurisdictions have leased or sanctioned marine locations for the exclusive use of such enterprises. Although the purchase and sale of fishing rights (and the inevitable consolidation of ownership that follows) have historically been viewed as inappropriate in the United States, similar strategies are widespread in the British Commonwealth countries. Private salmon fishing rights are common in English and Scottish rivers. Throughout the United Kingdom, an owner can voluntarily sell or lease rights. In Quebec, where the government has established exploitation zones managed by local associations on salmon rivers, overexploitation has declined and local contributions to enforcement and resource management have increased (Anderson and Leal, 1996). Native American communities in the Pacific Northwest once had nontransferable salmon fishing rights at particular locations on salmon rivers.

Private use rights for marine resources raises two questions. First, how feasible is establishing exclusive (community or individual) rights to harvest various marine species, and second, will it lead to better resource management? The following discussion deals with the options available for restricting access to marine resources by strengthening the rights of defined groups or individuals.

Community Access Rights

Exclusive, enforceable access rights can be assigned to particular groups, such as the residents of a particular community or a group that has traditionally used certain waters. Creating a sense of "ownership" over a resource may strengthen incentives for conserving it over time. However, the allocation of shares among members of the group remains a problem. Some of the 18 salmon fishery zones in Quebec have resolved this problem by establishing fishing fees (up to $44 per day) for members of the association and for nonmembers. The revenues are used to control poaching and for conservation projects (Anderson and Leal, 1996).

Informal (and sometimes illegal) community control over local lobster fisheries has also reduced overexploitation. Acheson's (1975) study of Maine lobster fisheries shows that the average size of lobsters tends to be larger where "harbor gangs" effectively exclude outsiders.

Individual Access Rights (Limited Access)

Access rights can be restricted to licensed individuals or entities (whose use of the resource may be further regulated). Eligibility may be defined in various ways, such as by completion of an apprenticeship program or by membership in a community. The allocation of limited rights can be carried out by various means, including "grandfathering" historical users or selling access rights. Aquaculture sites, for example, are leased to individuals and corporations in many coastal states.

License limitation programs are common in North America, Australia, and New Zealand. These programs attempt to control entry into a fishery and may facilitate cooperative management. But a license to fish, unless it also limits the catch, does not affect a fishermen's basic incentive to compete for fish. In fisheries where there are too many fishing vessels and technology is not controlled, licensing may not conserve stock or minimize costs.

Limitations on harvests are imposed in the United States, either by controlling the number of boats or by limiting the number of commercial licenses on a seasonal basis. For example, commercial salmon fishing permits are used to limit the total number of licenses disbursed in salmon fisheries in Bristol Bay, Alaska. In some cases, license limitations can be used as a moratorium on new users by limiting access to those already engaged in a harvest.

Individual Harvest or Use Rights

Licensed entities can be limited to certain amounts (e.g., catch quotas) or shares of the resources. Some traditional systems of managing naturally fluctuating resources (such as irrigation water in semi-arid climates) assign rights to given shares of the available resource. These arrangements reduce the incentive for overcapitalization but require an effective mechanism for allocating shares.

Quotas on individual fishermen promise to improve the economic efficiency of fisheries.[1] Quotas have been implemented in New Zealand's offshore and inshore fisheries (where they are called individual tradable quotas), in Australia's southern bluefin tuna and southeastern trawl fisheries, in a large number of Canada's freshwater and saltwater fisheries, and in the U.S. surf clam, wreckfish, ocean quahog, sablefish, and halibut fisheries. Privately owned individual quota shares (like water rights in western states) closely approximate exclusive property rights to the fish stock. Although owners do not own specific segments of the fish stock, they do have strong incentives to invest in the stock and to protect fish habitats. When quota shares are traded in competitive markets, the share prices approximate the economic value of fish (Anderson and Leal, 1996).

In New Zealand, the quota scheme has had mixed results. It has worked well with abalone beds, where fishermen have voluntarily stepped up security to stop poachers. But in the orange roughy fishery, the stock collapsed. Scientists determined that the breeding cycle of the fish was much longer than they had believed when the quotas were set. To reduce quotas and the resulting pressure on the fishery, the quotas had to be bought back by the government at great expense (Huppert, 1988; Annala, 1996).

Limiting Land-Based Growth

Point and nonpoint discharges into coastal waters, the volume of recreational use, and other demands on the marine environment are largely determined by the extent and pattern of land-based development in the coastal zone. Growth controls in the coastal zones are important tools for managing marine resources, which are otherwise the passive recipients of demands emanating on land. Growth can be controlled not only by controlling permitting for new construction but also by judiciously controlling public investments in infrastructure.

Control of land-based growth may be exercised through zoning restrictions, such as limiting population density or regulating coastal management and community master planning. Several federal and state laws include limitations on land-based growth. The federal CZMA (Coastal Zone Management Act) of 1972 encourages the creation of resource management plans to control growth and development in the coastal zone. Two acts in Washington state include limitations on land-based growth. The first, the Shoreline Management Act of 1971, attempts to balance resource use and resource protection with economic development and public access. This act mandated shoreline master programs to facilitate

[1]An NRC committee is currently conducting a study of individual fishing quotas and the more general question of rights-based allocation mechanisms. The report is expected to be available in late 1998.

View of Boston waterfront. Photo courtesy of William Eichbaum.

planning and permitting in an attempt to aid decision making and manage resources on a regional scale. The second act, the Growth Management Act of 1990, also attempts comprehensive regulation of development. Under this law, the state proactively assists localities grappling with decisions about limiting land-based growth.

Pricing Access

Demands on marine resources can also be rationed by pricing mechanisms, such as user charges and fees. The advantage of pricing mechanisms, is that they discourage uses with low economic values. Pricing mechanisms also reflect the true economic value of marine resources in commercial and recreational activities, values that would otherwise be treated as zero. Pricing marine resources creates incentives for users of the marine environment and marine resources to internalize the environmental costs (i.e., negative externalities) associated with their activities. Theoretically, if private costs equal social costs, the efficient use of resources will be encouraged.

Permit and license fees for boaters, commercial and recreational fishermen, tour and dive boat operators, waste dischargers, and other users have typically been low and cover only administrative costs. These fees could be used, however, to limit demand to some target level, but raising fees to limit demand will price

Industrial treatment lagoon, adjacent to Key Bridge, Baltimore Harbor, Maryland. Photo courtesy of William Eichbaum.

some users out of the market. Auctioning the rights to develop a resource to the highest bidder is an example of how a pricing mechanism can be used to limit demand.

User charges and fees intended to limit demand will increase public revenues from the resource. For some marine resources, such as offshore oil and gas, these revenues are considered to be the public's share of marine resources held by the government in the national interest. Other valuable marine resources, including fisheries and recreational services, such as whale watching operations, are now exploited by profit-seeking companies without substantial payment to the government. The general public is thereby denied its share of the value of the resource.

Revenues generated by user charges and fees can also be used to finance resource management and conservation in the coastal zone. Because general revenues are under severe pressure at the federal and state levels, using user fees as a source of revenue for high-priority expenditures should be given greater consideration.

User charges and related expenditures can also be useful as private sector management tools. For example, in a fishery with limits on group access, the problem of overcapitalization could be addressed by the fishery association financing a buyout program with fees levied on the members of the association. This would create a "win-win" situation for the association as a whole because the fishermen who finance the buyouts would benefit from the reduced capacity, and those who left the fishery would receive compensation. In addition, an industry-financed buyout, in contrast to a government-financed buyout financed from government contributions, would create stronger incentives for the association to ensure that buyouts reduced fishing capacity commensurate with expenditures (i.e., that their money was well spent.)

A successful buyout program was undertaken in Iceland, where salmon quotas held by commercial fishermen in Greenland and the Faroe Islands were bought out for three years by the National Fish and Wildlife Foundation. As a result, the numbers of salmon returning to rivers in Iceland and Europe doubled. Not only were stocks rebuilt, but the increase in inland sport fishing also gave Iceland a boost in employment and income (Anderson and Leal, 1996).

In 1996, the minister of fisheries and oceans of British Columbia implemented a license retirement program as part of a Pacific salmon revitalization plan with the goal of reducing the capacity of the West Coast commercial salmon fleet by 20 percent. The purpose of the program was to reduce the number of licenses in the salmon fleet equitably and quickly. Under the license retirement program, funds were made available to retire licenses. All salmon vessel owners holding full-fee and reduced-fee salmon licenses were eligible to apply. A Fleet Reduction Committee was set up to review all offers and to recommend to the Department of Fisheries and Oceans which licenses should be retired. A total of 800 commercial licenses were retired at an estimated cost of $80 million.

An additional restriction on salmon fishing is in effect in British Columbia, where the holder of a license now has access to only one area on the Pacific coast. Fishermen who wish to fish another area are required to purchase another license. Holding multiple licenses, known as "license stacking," is, in essence, a voluntary fleet reduction or an industry-financed buy-back program.

"Cap and Trade" Mechanisms

An increasingly popular instrument of environmental policy involves determining an allowable ceiling on the use of a resource and enabling users to trade allowances among themselves. For example, total effluent limits have been established for different classes of pollutants, and emitters have been permitted to trade quanta of emissions among themselves. However, quotas in fisheries are more difficult to maintain than to impose so the outcome is not always clear.

Similar mechanisms have been used to limit development in ecologically sensitive areas by requiring prospective developers (beyond a predetermined scale) to purchase development rights from other landholders. Cap and trade mechanisms could be used (like taxicab medallions) to limit the number of commercial tour and dive operators in ecologically sensitive areas or (like pollution trading) to limit the amount of point-source effluent discharged into coastal waters. Cap and trade mechanisms introduce flexibility and incentives for efficiency into regulatory systems, but, of course, they do not resolve the basic problem of establishing appropriate limits.

Enforcement Problem

Many activities that degrade the marine environment, such as the illegal harvesting of commercial species by foreign vessels and the disposal of vessel waste at sea, occur out of sight of most observers. Because few people are on the water to see what goes on, violations of regulations are difficult to detect. Although the majority of firms and individuals working in the marine environment are law-abiding, the minority creates an enforcement problem. However, there are management tools that can make enforcement easier.

Improving Monitoring

Recent technological advances in monitoring have created the potential for more accurate and comprehensive tracking and identification of marine activities. Observational satellites and global positioning systems can track ships far from shore. Previous violators of navigational, dumping, or fishing regulations might be required to carry transponders that would allow remote monitoring of their movements. Chemical tracers and "fingerprints" have been developed that might enable analysts to determine the source of ocean spills. New monitoring technologies could be employed more vigorously to detect and, thus deter, infractions of marine regulations.

More Severe Sanctions

Chronic violators of marine regulations are acting on reasoned expectations of likely gains and losses from their transgressions. Expectations of losses are based on the probability of their activities being detected and the penalty they might face. It follows that when the probability of detection is low, effective deterrence requires that the penalties be onerous. But penalty schedules do not always conform to this model. Sometimes they amount to little more than giving up illicit gains, which has little, if any, deterrent effect. Heavier penalties, especially for repeat violators, would probably improve enforcement.

Involving the Community in Rule-Making and Enforcement

A participatory approach to rule-making increases the likelihood that those to whom the rules apply will perceive them as legitimate and also ensures that rules are appropriate to local conditions. Community enforcement, enhanced by local knowledge and peer pressure, has been successful in regulating local fisheries. However, a broad "community" that includes all relevant stakeholders must be involved, not just stakeholders who exploit the resource commercially.

In land-based enforcement, broad participation has been encouraged through various inducements to "whistle blowers," including sharing fines or penalties with whoever reports and documents violations. This kind of enforcement mechanism seems applicable to marine area management as well.

Financing Marine Area Governance and Management Programs

Improving marine area governance will undoubtedly involve significant costs. There are mechanisms, however, that could be used to generate funds to cover programmatic costs. Traditional financing mechanisms that could potentially be applied to marine management programs include bonds, taxes, and grants and loans (see Appendix D for a discussion of these options).

Note that the use of bonds and taxes may require new legal mandates, and given the current political climate, may not be feasible. In addition, financing management programs solely from federal and state taxes, grants, and bond issues is becoming increasingly difficult as pressures on government budgets increase. Therefore, financing improved marine management programs will require innovative financing approaches (EPA, 1988; Kearney, 1994). Two such approaches are described below. (See Appendix D for a more comprehensive list.)

Chesapeake Bay Sports Fishing License Program

In response to deteriorating water quality in the Chesapeake Bay, the state of Maryland began a five-point program to improve water quality and manage the abundant natural resources of the bay. As part of this program, the state instituted the Chesapeake Bay Sport Fishing License plan in January 1985 and became the first East Coast state to license tidal water anglers. Fees collected from sport fishing licenses are credited to the Fisheries Research and Development Fund and are used to propagate and conserve native fish stocks. The ultimate goal of the program is to improve sport fishing and to support research on tidal fishery resources. Fees on sport fishing licenses generate considerable revenues for estuarine and marine management, depending on the strength of the regional sport fishing industry.

Under this program, no one is allowed to fish in the Chesapeake Bay or its tributaries up to the tidal boundaries without first obtaining a Chesapeake Bay Sport Fishing License.[2] In addition to the basic license, special licenses must be obtained for charter boats or senior citizens. Box 6-1 outlines the different types of licenses, the number of licenses sold, and the total revenue from the program. The program is overseen by the Maryland Department of Natural Resources, and

---

[2]Exceptions include holders of Virginia Chesapeake Bay fishing licenses, commercial fishermen, and children under 16. Note: a $2.00 striped bass stamp is required of everyone.

Chesapeake Bay. Photo courtesy of William Eichbaum.

anyone caught fishing without a license is penalized. The enforcement officer generally issues a warning for the first offense.

Fees for sport fishing licenses can be used to fund management of the marine fish stock and sport fishing bases. License fees used to generate revenue for estuarine and marine management could be extended to other recreational activities, such as boating. Revenues from licensing fees could be used as seed money for revolving loan funds with the proceeds dedicated to marine area management.

Clean Water Districts[3] in Washington State

In 1992, the Washington State legislature passed a provision for the creation of shellfish protection districts—more commonly referred to as clean water districts (CWDs)—to prevent the contamination of commercial and recreational shellfish beds and to restore water quality in areas already affected by nonpoint source pollution. Shellfish protection districts provide a mechanism for generating funds for improving or maintaining water quality. CWDs can be created by a county legislative authority or by voter referendum. If the State Department of Health has issued a downgrade or closure of a shellfish growing area because of nonpoint source pollution, counties in the downgrade area are required to establish a CWD within 180 days. District boundaries may cover an individual

---

[3]Also called shellfish protection districts.

watershed, an entire county, or, by interjurisdictional agreement, parts of several counties and incorporated areas. Seven CWDs have been established to date.

Once a CWD has been established, a citizens advisory committee determines priorities for controlling pollution. Counties finance CWD programs through taxes, fees, rates, charges for specified protection programs, and grants or loans from other sources. The combination of revenue sources is determined by the county legislative authority.

In Mason County, for example, property owners in the Lower Hood Canal CWD are assessed $52 per year for structures with on-site septic systems. The annual fee for complexes with multiple connections to a septic system is $250; the fee for state parks is $450. Tideland property owners are assessed $26 per year. The fees are supplemented by state grants (some of which require a 25 percent local match), which are dedicated to nonpoint source pollution control.

CWDs are an example of a mechanism that funds comprehensive water pollution management at the local level. This mechanism could be modified to deal with other marine area management problems, dredging, and dredged spoils disposal.

BOX 6-1
Fishing License Sales by the Maryland Department of Natural Resources, 1996

Fishing License and Stamps	Quantities Sold	Revenue ($$)
Resident Nontidal	168,001	$1,602,121
Senior Resident Consolidated	18,290	83,142
Nonresident Nontidal	14,093	269,394
Five-Day Resident Nontidal	1,489	8,634
Five-Day Nonresident Nontidal	10,738	69,381
Trout Stamps	69,203	344,352
Replacement Nontidal	382	393
Resident Bay Sport	147,228	957,207
Nonresident Bay Sport	24,009	276,055
Five-Day Resident Bay Sport	3,584	12,587
Five-Day Nonresident Bay Sport	15,988	56,025
Pleasure Boat Decal	32,471	962,251
Charterboat 6	335	80,450
Charterboat 7	88	25,420
Replacement Bay Sport	281	759
Replacement Charterboat 6/7	8	40
Totals	506,188	$4,749,191

Source: Maryland Department of Natural Resources.

SUMMARY

Improving marine area governance necessarily requires confronting fundamental underlying problems, such as the prevalence of externalities, open access to marine resources, and the unrestrained increase in demand for resources in the public domain. Whatever the institutional arrangements, responsible bodies must address these problems with effective management tools and approaches. Direct regulation has proven to be cumbersome and often ineffective. The benefits of other approaches, especially approaches that attempt to reconcile private economic incentives with the overall objectives of resource management, have not yet been fully realized.

7

Conclusions and Recommendations

INTRODUCTION

The governance and management of our coastal waters are inefficient and wasteful of both natural and economic resources. The existing system is characterized by a confusing array of laws, regulations, and practices at the federal, state, and local levels, and agencies that implement and enforce existing systems operate with mandates that often conflict with each other. No mechanism exists for establishing a common vision and a common set of objectives.

Government agencies operate in an arena characterized by unresolved conflicts among values and economic expectations, which require difficult choices among competing needs and interests and raise substantial questions of equity both with regard to present interests and the interests and rights of future generations. These conflicts impose a number of costs on advancing the national interest, both direct and indirect, such as lost opportunities and economic costs. The environmental costs of some conflicts are not always readily apparent. For example, the cumulative impacts of development can result in substantial alteration of a habitat that may only be revealed long after the development has ended.

Resolving conflicts necessarily involves many difficult decisions by public agencies about the appropriate balance between using resources to satisfy immediate economic needs and preserving resources for future needs. Under the present system, decisions are made on a case-by-case basis, often involving costly and lengthy processes. No coherent system to protect the overall national interest is available to guide decision-makers.

Establishing basic principles and effective processes for the governance of the ocean and coastal areas is a prerequisite both to economic investment and to sound environmental stewardship and would make a more reasonable, less adversarial approach to resolving conflicts possible. The general elements of a framework improved for governance and management envisioned in this report include the following:

- Goals must be clearly stated, especially when different entities must be brought together in a cooperative effort.
- The geographic (or ecological) area to be managed needs to be carefully delineated.
- Mechanisms for involving all relevant stakeholders in the governance process need to be designed.
- In most situations, it is appropriate for the process to be initiated as a state-federal joint effort.
- Systems should foster innovative responses to management and resource utilization.
- Processes should be established for incorporating scientific information into all aspects of decision-making.
- Success should be measured by monitoring and evaluation.

A fully developed marine governance and management system that meets all of the objectives and incorporates all of the elements discussed in this report must evolve over time and in response to actual experience. However, some measures can be taken now.

The following recommendations are intended to bring the strengths inherent in the federalist approach, which is well developed in the management of public lands, to the marine governance system. In the federalist system, states and local governments are granted equal partnership with the federal government as appropriate.

IMPROVING GOVERNANCE

Conclusion 1. The lack of coordination in the marine governance system diminishes the effectiveness of agencies at all levels and results in the loss of economic and ecological opportunities.

Recommendation 1. A National Marine Council should be established to define national objectives in the marine environment and to coordinate the activities of federal agencies, state agencies, and interested parties in the private sector.

Conclusion 2. The governance and management of ocean uses and resources are poorly coordinated at the regional level and often fail to involve nongovernmental parties in decision making.

Recommendation 2. Regional councils authorized by the National Marine Council should be created where there are serious conflicts or high resource values and existing programs are not available or are not effective. Regional councils can provide technical assistance on marine management issues, facilitate the use of scientific and monitoring information, develop alternative processes for resolving disputes, facilitate participation by local interests in governance decisions, and pursue contractual arrangements with stakeholders and other participants to achieve management goals.

Conclusion 2. Although many federal and state programs exist, no integrated, coherent overall structure for marine governance and management has been established.

Recommendation 3. Federal officials, working with their state counterparts, should attempt to maximize existing programs, especially where there are urgent problems and existing programs could be reconfigured relatively easily to provide some, or all, of the benefits associated with regional councils.

IMPROVING MANAGEMENT

Conclusion 4. A wide range of management tools have been adopted for land management but have not been used much in the marine environment.

Recommendation 4. Management tools should be explored and adapted as needed to improve marine governance, both by the proposed regional marine councils and by existing marine management programs. Management tools include zoning, enhanced systems of liability or compensation for economic and environmental damage, user charges and marketable use rights, and negotiating the mitigation of activities that are potentially harmful to other resource users and values.

Conclusion 5. The marine environment presents special difficulties for devising and implementing governance processes because of the tradition of open access that has characterized marine resources and space.

Recommendation 5. In appropriate situations, limiting access by creating alternative rights, such as community access rights, controlled access, or individual use rights, should be considered.

Conclusion 6. Many goods and services provided by the marine environment are considered to be of no economic value because they are not traded in an economic market. Nevertheless, the value of these services may greatly exceed their commercial value.

Recommendation 6. Management agencies should make every effort to estimate the value of nonmarketable marine services, such as recreation and ecosystem stability, and should reflect those values in management decisions and decision making.

Conclusion 7. Many activities that degrade marine resources and areas take place out of sight or over extended periods of time and are, therefore, not easy to document.

Recommendation 7. The federal government should ensure compliance with legal requirements by improving surveillance, strengthening sanctions, and involving all elements of the marine community in more transparent rule-making and enforcement in marine areas.

Conclusion 8. Marine governance has been hampered by inadequate financial resources.

Recommendation 8. A wide range of financing mechanisms that are now used on land should be considered for the marine environment. These include performance bonds, use or resource-based taxes, grants and loans, special assessment districts, recovery of costs for government services, tax-increment financing.

Conclusion 9. The effectiveness of existing programs could be enhanced by a broader range of management tools for dealing with problems and conflicts.

Recommendation 9. Existing federal and state coastal and marine programs should examine and, where appropriate, adopt new governance mechanisms and management tools that foster coordination and cooperation.

IMPLEMENTING CHANGE

Although the implementation of the system of marine area governance and management proposed in this report would involve creating new structures and processes, a number of the measures suggested in this report can be taken immediately by the federal and state agencies responsible for marine and coastal activities and areas. The most important components of the system discussed in the preceding pages are coordination, information, and participation. These principles can be integrated into existing or new management or governance frameworks and would yield immediate, substantial benefits to the nation.

References

Acheson, J.M. 1975. The lobster fiefs: economic and ecological effects of territoriality in the Maine lobster industry. Human Ecology 3(3): 183–207.

Anderson, T., and D. Leal. 1996. The Salmon in Economics. Bozeman, Mont.: Political Economy Research Center.

Annala, J.H. 1996. New Zealand's ITQ system: have the first eight years been a success or a failure? Reviews in Fish Biology and Fisheries 6: 43–62.

Bacon, C. 1996. Time Trend Data for U.S. Coastal Resources. Unpublished paper available from the Marine Board, National Research Council, Washington, D.C.

Becker, C.D., and E. Ostrom. 1995. Human ecology and resource sustainability: the importance of institutional diversity. Annual Review of Ecological Systems 26: 113–133.

Bell, F.W. 1972. Technological externalities and common property resources: an empirical study of the U.S. northern lobster fishery. Journal of Political Economy 80(1): 148–158.

Bell, F.W. 1989. Application of Wetland Evaluation Theory to Florida Fisheries. Tallahassee, Fla.: The Florida Sea Grant College.

Bigelow, J. 1992. Developing managerial wisdom. Journal of Management Inquiry 1(2): 143–153.

Bingham, G. 1995. Issues in ecosystem valuation: improving information for decision making. Ecological Economics 14: 73–90.

Binswanger, H., M. Faber, and R. Manstetten. 1990. The dilemma of modern man and nature: an exploration of the Faustian imperative. Ecological Economics 2: 197–223.

Bockstael, N., R. Costanza, I. Strand, W. Boynton, K. Bell, and L. Wainger. 1995. Ecological economic modeling and valuation of ecosystems. Ecological Economics 14(2): 143–159.

Botkin, D.B. 1990. Discordant Harmonies. New York: Oxford University Press.

Bowman, A. and R. Kearney. 1986. The Resurgence of the States. Englewood Cliffs, N.J.: Prentice-Hall.

Boynton, W.R., J.H. Garber, R. Summers, and W.M. Kemp. 1995. Inputs, transformations, and transport of nitrogen and phosphorus in Chesapeake Bay and selected tributaries. Estuaries 18: 285–314.

Butts, R. 1988. Management of the Marine and Ocean Resources of the Washington Coast: An Interim Report to the Washington State Legislature. Joint Select Committee on Marine and Ocean Resources. Olympia, Wash.: Washington State Legislature.

Capuzzo, J.M. 1995. Environmental Indicators of Toxic Chemical Contaminants in the Gulf of Maine. Pp. 187–204 in Improving Interactions Between Coastal Science and Policy: Proceedings of the Gulf of Maine Symposium. November 1–3, 1994. Kennebunkport, Maine. National Research Council. Washington, D.C.: National Academy Press.

Catena, J. 1992. Policy Options for Maine's Marine Waters: A Report of the Marine Policy Committee of the Land and Water Resources Council. Augusta, Me.: Maine Coastal Program, State Planning Office.

Center for Urban and Regional Studies. 1991. Evaluation of the National Coastal Zone Management Program. Chapel Hill, N.C.: University of North Carolina Press.

Checkland, P. 1982. Systems Thinking, Systems Practice. New York: John Wiley & Sons.

Christie, D. 1989. Florida's Ocean Future: Toward a State Ocean Policy. Tallahasse, Fla.: Environmental Policy Unit Report for the Governor's Office of Planning and Budgeting.

Christensen, J.P., D.B. Smith, and L.M. Mayer. 1992. The nitrogen budget of the Gulf of Maine and climate change. Pp. 75–90 in Proceedings of the Gulf of Maine Scientific Workshop. Wiggin and Mooers (eds.) January 8–10, 1991, Woods Hole, Massachusetts. Boston: Urban Harbors Institute, University of Massachusetts-Boston.

Cicin-Sain, B., M. Hershman, R. Hildreth, and J. Isaacs. 1990. Improving Ocean Management Capacity in the Pacific Coast Region: State and Regional Perspectives. Newport, Ore.: National Coastal Resources and Development Institute.

Cicin-Sain, B., and R. Knecht. 1985. The problem of governance of U.S. ocean resources and the new exclusive economic zone. Ocean Development and International Law 15: 289.

Clark, D.L. 1985. Emerging paradigms in organizational theory and research. Pp. 43–78 in Y.S. Lincoln (ed.), Organizational Theory and Inquiry. Newbury Park, Calif.: SAGE Publications.

Clemson, B. 1984. Cybernetics: A New Management Tool. Turnbridge Wells, Kent, UK: Abacus Press.

Costanza, R., R. d'Arge, R. deGroot, S. Farber, M. Grasso, B. Hannon, K. Limberg, S. Naeem, R.V. O'Neill, J. Paruelo, R.G. Raskin, P. Sutton, and M. van den Belt. 1997. The value of the world's ecosystem services and natural capital. Nature 387: 253–260.

Council on Environmental Quality. 1992. United States of America National Report. Washington, D.C.: U.S. Government Printing Office.

Culliton, T.J., M.A. Warren, T.R. Goodspeed, D.G. Remer, C.M. Blackwell, and J. MacDonough. 1990. Fifty years of Population Change along the Nation's Coast. Coastal Trends Series. Washington, D.C.: National Oceanic and Atmospheric Administration.

Davenport, T.H., R.G. Eccles, and L. Prusak. 1992. Information politics. Sloan Management Review 34(1): 53–65.

Eichbaum, W.E. 1984. Cleaning up the Chesapeake Bay. Environmental Law Reporter 14(6): 10237–10245.

Eichenberg, T., and J. Archer. 1987. The federal consistency doctrine: coastal zone management and the "new federalism." Ecology Law Quarterly 14: 9.

Eisenberg, E.M. 1990. Jamming: transcendence through organizing. Communications Research 17(2): 139–164.

Environmental Protection Agency (EPA). 1988. Financing Marine and Estuarine Programs: A Guide to Resources. Report no. 503/8–88/001. Washington, D.C.: U.S. EPA.

EPA. 1995. The State of the Chesapeake Bay, 1995. Washington, D.C.: U.S. EPA.

Fitzgerald, R. 1987. Exxon v. Fischer: thresher sharks protect the coastal zone. British Columbia Environmental Affairs Law Review 14: 561.

Fogerty, J.J., M.P. Sissenwine, and E.B. Cohen. 1991. Recruitment variability and the dynamics of exploited marine populations. Trends in Ecology and Evolution 6: 241–246.

Freeman, A.M. III. 1995. The benefits of water quality improvements for marine recreation: a review of the empirical evidence. Marine Resource Economics 10(4): 385–406.

Good, J.W., and R.G. Hildreth. 1987. Oregon Territorial Sea Management Study. Salem, Ore.: Oregon Department of Land Conservation and Development.

Hackman, J.R., ed. 1990. Groups That Work (and Those That Don't). San Francisco: Jossey-Bass.

Hardin, G. 1968. The tragedy of the commons. Science 162: 1243–1248.

Hawaii Ocean and Marine Resources Council. 1988. Hawaii Ocean Resources Management Plan. Honolulu: State of Hawaii.

Hedgpeth, J.W. 1993. Foreign invaders. Science 261: 34–35.

Hershman, M., D. Fluharty, and S. Powell. 1988. State and Local Influence Over Offshore Oil Decisions. Seattle: Washington Sea Grant Program, University of Washington.

Heydebrand, W.V. 1989. New organizational forms. Work and Occupations 16(3): 323–357.

Hildreth, R. 1995. Ocean Planning by U.S. Coastal States: Legal Implications for Ocean Resource Use and Preservation. Seattle: University of Washington School of Law, Washington Law School Foundation, Continuing Legal Education.

Hoagland, P., H.L. Kite-Powell, and M.E. Schumacher. 1996. Marine Area Governance and Management in the Gulf of Maine: A Case Study. Unpublished paper available from the Marine Board, National Research Council, Washington, D.C.

Huppert, D.D. 1988. The Role of Economic Analysis in Marine Fishery Regulation. Pp. 88 in Marine Fishery Allocations and Economic Analysis: Proceedings of a Regional Workshop. J.W. Milon (ed). May 1988. Gainesville, Fla.: University of Florida Press.

Hutchins, E. 1990. The technology of team navigation. Pp. 191–220 in Intellectual Teamwork. J. Galegher, R.E. Kraut, and C. Egido (eds.). Hillsdale, N.J.: Erlbaum.

Imperial, M.T., and T.M. Hennessey. 1996. An ecosystem–based approach to managing estuaries: an assessment of the National Estuary Program. Coastal Management 24(2): 115–139.

Isenberg, D.J. 1985. Some hows and whats of managerial thinking. Pp. 168–181 in J.G. Hunt and J.D. Blair, (eds.). Leadership on the Future Battlefield. Washington, D.C.: Pergamon-Brassey.

Janowitz, M. 1959. Changing patterns of organizational authority: the military establishment. Administrative Science Quarterly 3: 473–493.

John, D. 1994. Civic Environmentalism: Alternatives to Regulation in States and Communities. Aspen Institute and National Academy of Public Administration, Washington, D.C.: Congressional Quarterly Press.

John, D. 1996. Protecting a Profitable Paradise: The National Ocean Service Leads Multi-Agency Planning in the Florida Keys. Unpublished paper available from the Marine Board, National Research Council, Washington, D.C.

Kalberg, S. 1980. Max Weber's types of rationality: cornerstones for the analysis of rationalization processes in history. American Journal of Sociology 85: 1145–1179.

Kaoru, Y., V.K. Smith, and J.L. Liu. 1995. Using random utility models to estimate the recreational value of estuarine resources. American Journal of Agricultural Economics 77(1): 141–151.

Kearney, A.T., Inc. 1994. A Framework for Identifying Financing Approaches for Implementing Comprehensive Conservation and Management Plans Developed Under the National Estuary Program. Unpublished report. U.S. Environment Protection Agency, Washington, D.C.

Kelleher, G., C. Bleakley, and S. Wells. 1995. A Global Representative System of Marine Protected Areas, Vol. 1. The Great Barrier Reef Park Authority. London, U.K.: The World Bank and the World Conservation Union.

Kelly, E. 1988. Zoning. Pp. 251–286 in The Practice of Local Government Planning (2nd ed.). Washington, D.C.: International City Management Association.

King, J., and B. Olson. 1988. Coastal state capacity for marine resource management. Coastal Management 16: 305.

Kitsos, T. 1994. Troubled Waters: A Half Dozen Reasons Why the Federal Offshore Oil and Gas Program is Failing—A Political Analysis. Prepared for the 3rd Annual Symposium of the Ocean Governance Study Group. Lewes, Delaware.

Lester, R. 1994. Rediscovering the Public Interest in the Outer Continental Shelf Lands. Prepared for the 3rd Annual Symposium of the Ocean Governance Study Group. Lewes, Delaware.

Lynn, G.D., P. Conroy, and F.S. Prochaska. 1981. Economic valuation of marsh areas for marine production processes. Journal of Environmental Economics and Management 8(2): 175–186.

March, J.G., and H.A. Simon. 1958. Organizations. New York: John Wiley & Sons.

Marine Board. 1993. Marine Board Forum: The Future of the U.S. Exclusive Economic Zone. Marine Board, National Academy of Sciences. Unpublished proceedings. April 10, 1993.

Martin, D.M., T. Morton, T. Dobrzynski, and B. Valentine. 1996. Estuaries on the edge: the vital link between land and sea. Washington, D.C.: American Oceans Campaign.

McLaughlin, R., and L. Howorth. 1991. Mississippi Ocean Policy Study. University, Miss.: Mississippi-Alabama Sea Grant Legal Program, University of Mississippi Law Center.

Miles, R.E., and C.S. Snow. 1992. Causes of failure in network organizations. California Management Review 34(4): 53–72.

Mintzberg, H. 1983. Structure in Fives: Designing Effective Organizations. Englewood Cliffs, N.J.: Prentice-Hall.

Mitchell, F. 1991. Preservation of state and federal authority under OPA. Environmental Law 21: 237.

Murawski, S.A. 1996. St. Peter's thumbprint: An industrial-ecological history of the New England groundfish fishery in the 20th century. Woods Hole, Mass.: Woods Hole Laboratory, New England Fisheries Science Center, National Marine Fisheries Service.

National Oceanic and Atmospheric Administration (NOAA). 1996. Florida Keys National Marine Sanctuary: Strategy for Stewardship. Final Management Plan/Environmental Impact Statement, Vol I–III. Washington, D.C.: U.S. Department of Commerce.

NOAA. 1997. NOAA Fisheries Strategic Plan. Washington, D.C.: U.S. Department of Commerce.

National Performance Review. 1993. From Red Tape to Results: Creating a Government that Works Better and Costs Less. Washington, D.C.: U.S. Government Printing Office.

National Research Council (NRC). 1989. Our Seabed Frontier. Washington, D.C.: National Academy Press.

NRC. 1990a. Interim Report of the Committee on Exclusive Economic Zone Information Needs: Coastal States and Territories. Washington, D.C.: National Academy Press.

NRC. 1990b. Managing Troubled Waters: The Role of Marine Environmental Monitoring. Washington, D.C.: National Academy Press.

NRC. 1991. Interim Report of the Committee on Exclusive Economic Zone Information Needs: Seabed Information Needs of Offshore Industries. Washington, D.C.: National Academy Press.

NRC. 1992. Working Together in the EEZ: Final Report of the Committee on Exclusive Economic Zone Information Needs. Washington, D.C.: National Academy Press.

NRC. 1994a. Improving the Management of U.S. Marine Fisheries. Washington, D.C.: National Academy Press.

NRC. 1994b. Priorities for Coastal Ecosystem Science. Washington, D.C.: National Academy Press.

NRC. 1994c. Restoring and Protecting Marine Habitat: The Role of Engineering and Technology. Washington, D.C.: National Academy Press.

NRC. 1995a. Beach Nourishment and Protection. Washington, D.C.: National Academy Press.

NRC. 1995b. Improving Interactions Between Coastal Science and Policy: Proceedings of the Gulf of Maine Symposium. Kennebunkport, Maine. November 1-3, 1994. Washington, D.C.: National Academy Press.

NRC. 1995c. Science, Policy, and the Coast. Washington, D.C.: National Academy Press.

NRC. 1995d. Understanding Marine Biodiversity. Washington, D.C.: National Academy Press.

NRC. 1995e. Wetlands: Characteristics and Boundaries. Washington, D.C: National Academy Press.

Natural Resources Defense Council (NRDC). 1996 Testing the Waters: Who Knows What You're getting into. New York: NRDC publications.

NRDC. 1997. Hook, Line, and Sinking: The Crisis in Marine Fisheries. New York: NRDC Publications.

Norse, E.A. 1993. Global Marine Biological Diversity: A Strategy for Building Conservation into Decision Making. Washington, D.C.: Island Press.

North Carolina Marine Science Council. 1984. North Carolina and the Sea: An Ocean Policy Analysis. Raleigh, N.C.: Office of Marine Affairs, North Carolina Department of Administration.

Office of Management and Budget. 1994. Catalog Domestic Assistance. Library of Congress No. 73–600118. Washington D.C.: Government Printing Office.

Osborne, R. 1988. Laboratories of Democracy. Boston: Harvard Business School Press.

Outer Continental Shelf Policy Committee. 1993. Moving beyond Conflict to Consensus. Report of the OCS Policy Committee's Subcommittee on OCS Legislation. Washington, D.C.: U.S. Department of the Interior.

President's Council on Sustainable Development. 1996. Sustainable America: A New Consensus for Prosperity, Opportunity, and a Healthy Environment for the Future. Washington, D.C.: U.S. Government Printing Office.

Roberts, K.H., ed. 1993. New Challenges to Understanding Organizations. New York: Macmillan.

Roberts, K.H., S.K. Stout, and J.J. Halpern. 1994. Decision dynamics in two high reliability military organizations. Management Science 40(5): 614–624.

Rochlin, G.I. 1989. Informal organizational networking as a crisis avoidance strategy: U.S. naval flight operations as a case study. Industrial Crisis Quarterly 3: 159–176.

Sahl, J.D., and B.B. Bernstein. 1995. Developing policy in an uncertain world. International Journal of Sustainable Development World Ecology 2: 124–135.

Simon, H.A. 1957. Administrative Behavior (2nd ed.). New York: Macmillan.

Smith, V.K., and Y. Kaoru. 1990. Signals or noise? explaining the variation in recreation benefit estimates. American Agricultural Economics Association 72(2): 419–433.

Stroud, R.H. 1994. Conserving America's Fisheries. Proceedings of a National Symposium on the Magnuson Act, New Orleans, March 8–10, 1993. Savannah, Ga: National Coalition for Marine Conservation.

United Nations Commission on Environment and Development (Brundtland Commission). 1987. Our Common Future. Oxford, U.K.: Oxford University Press.

U. S. Department of Natural Resources. 1995. Restoring the Chesapeake Bay: Chesapeake Bay Progress Report. Annapolis, Md.: State of Maryland.

Walsh, R.G., D.M. Johnson, and J.R. McKean. 1992. Benefit transfer of outdoor recreation demand studies, 1968–1988. Water Resources Research 28(3): 707–713.

Waterman, M. 1995. Marine protected areas in the Gulf of Maine. Natural Areas Journal 15: 43–49.

Weber, M. 1947. The Theory of Social and Economic Organization. London: Oxford University Press.

Weber, M.L. 1995. Healthy Coasts, Healthy Economy: A National Overview of America's Coasts. Washington, D.C.: Coast Alliance.

Weber, M.L., and J.A. Gradwohl. 1995. The Wealth of Oceans. New York: W.W. Norton.

Weick, K.E. Sources of order in underorganized systems: Themes in recent organization theory. Pp 103–136 in Y.S. Lincoln (ed.) Organizational Theory and Inquiry. Newbury Park, Calif.: SAGE Publications.

Weick, K.E., and K.H. Roberts. 1993. Collective mind in organizations: heedful interrelating on flight decks. Administrative Science Quarterly 38: 357–381.

Wilson, J.A., J.M. Acheson, M. Metcalfe, and P. Kleban. 1994. Chaos, complexity and community management of fisheries. Marine Policy 18(4): 291–305.

Zey, M. 1992. Criticisms of rational choice models. Pp. 9–31 in Decision Making: Alternatives to Rational Choice Models. M. Zey (ed). Newbury Park, Calif.: SAGE Publications.

Acronyms

CCMP	comprehensive conservation and management plan
CERCLA	Comprehensive Environmental Response, Compensation and Liability Act
CWD	clean water district
CZM	coastal zone management
CZMA	Coastal Zone Management Act
EEZ	exclusive economic zone
EPA	Environmental Protection Agency
FCMA	Fisheries Conservation and Management Act
GNP	gross national product
IFQ	individual fishing quota
MCPA	marine and coastal protected areas
MMS	Minerals Management Service
NAPA	National Academy of Public Administration
NEP	National Estuary Program
NEPA	National Environmental Policy Act
NMFS	National Marine Fisheries Service
NMS	national marine sanctuary
NOAA	National Oceanic and Atmospheric Administration

NOS	National Ocean Service
NRC	National Research Council
OCS	outer continental shelf
OCSLA	Outer Continental Shelf Lands Act
OPA90	Oil Pollution Act of 1990
SMBRP	Santa Monica Bay Restoration Project
TAC	total allowable catch
WHOI	Woods Hole Oceanographic Institution

APPENDICES

APPENDIX
A

Biographical Sketches of
Committee Members

William M. Eichbaum, *chair*, an environmental lawyer, is vice president of the World Wildlife Fund. He has been the undersecretary, Executive Office of Environmental Affairs, for the Commonwealth of Massachusetts and was assistant secretary for environmental programs for the Maryland Department of Health and Mental Hygiene. Mr. Eichbaum also served as general counsel and deputy secretary for the Pennsylvania Department of Environmental Resources and associate solicitor for surface mining for the U.S. Department of the Interior. Mr. Eichbaum received a B.A. from Dartmouth College and an L.L.B. from Harvard Law School and has published numerous articles on environmental law. He served on the Marine Board, the Marine Board Committee on Marine Environmental Monitoring, on the Water Science and Technology Board Committee on Coastal Waste Water Management, the Polar Research Board Antarctic Environmental Committee, and is currently serving on the Water Science and Technology Board.

Edward P. (Ted) Ames has engaged for more than 15 years in commercial fishing of a variety of species off the coast of Maine, including groundfish, pelagic fish, lobster, scallops, and sea urchins. He has also been president and laboratory director of Alden/Ames Laboratory, a high school and university science teacher, and marine resources director at the Island Institute. Mr. Ames serves on many committees and commissions, including the Maine Groundfish Hatchery Commission and the New England Fishery Management Council, and is executive director and president of the Maine Gilnetters Association. Mr. Ames has provided testimony and position papers related to state and federal fisheries legislation and is a frequent speaker at fisheries workshops and conferences. He has B.S. and M.S. degrees in biochemistry from the University of Maine at Orono.

Robert L. Bendick, Jr., is currently director of the Florida chapter of the Nature Conservancy. Prior to that, he was project manager at the New York State Department of Environmental Conservation, where he served as deputy commissioner for natural resources from 1990 to 1995. He was responsible for administration of the New York natural resource programs, including programs within the divisions of Fish and Wildlife, Lands and Forests, and Marine Resources. Mr. Bendick designed and implemented New York State's first open space conservation plan and supervised preparation of the first use and information plan for the Adirondack Forest Preserve. He served as chair of the Northern Forest Lands Council and prepared the comprehensive conservation and management plan for the Long Island Sound. Mr. Bendick previously served as director of the Rhode Island Department of Environmental Management, a cabinet level position responsible for all resource management and environmental protection activities in the state. Mr. Bendick has a bachelor's degree in history from Williams College and a master's degree in Urban Planning from New York University Graduate School of Public Administration.

Brock B. Bernstein is a partner of EcoAnalysis, Inc., a consulting firm specializing in database systems design, data management, and data analysis of environmental, fisheries, and marine biological research. He has more than 20 years of experience in marine research and environmental studies. He holds a Ph.D. in biological oceanography from Scripps Institution of Oceanography. A special research interest of Dr. Bernstein is the application of statistical and experimental design principles to environmental projects carried out under real-world constraints.

Leo R. Brien (deceased) served as maritime director for the Port of Oakland from October 1993 to June 1997. Mr. Brien joined the Port of Oakland after five years as president of the Pacific Merchant Shipping Association, a legislative and regulatory advocacy group for West Coast shipping lines. While there, he sponsored legislative initiatives on behalf of the industry. He was a founder of the Bay Dredging Action Coalition, an alliance of labor, management, and other interest groups formed to shape public policy in support of dredging. From 1973 to 1988, Mr. Brien held positions of increasing responsibility with Sea-Land Service, Inc., the largest U.S.-flag container line. He was on the Governor's Technical Advisory Committee and was president of the board of the Oakland Apostleship of the Seafarers' mission. Mr. Brien was a graduate of Boston College.

Charles G. (Chip) Groat is the executive director of the Center for Coastal, Energy, and Environmental Resources (CCEER) at Louisiana State University. CCEER encompasses 12 research institutions and organizations, including three graduate programs in oceanography and coastal science, environmental studies, and nuclear science. Dr. Groat previously was director and state geologist of the Louisiana Geological Survey and later served as assistant to the secretary for the Louisiana Department of Natural Resources, where he administered the state Coastal Zone Management Program and the Coastal Protection Program. He has taught university courses on geomorphology, energy and mineral resources, and the environmental aspects of resource development and has published numerous papers dealing with energy and mineral resources, water resources and quality, coastal geology, and resources and environmental policy. Dr. Groat received his Ph.D. in geology from the University of Texas at Austin.

Marc J. Hershman is professor of marine studies, adjunct professor of law, and director of the School of Marine Affairs at the University of Washington. He is the author and editor of books on coastal zone management, urban ports, and maritime history, as well as numerous publications dealing with law and policy affecting coastal and marine resources. He has been editor of the journal, Coastal Management, for 15 years and is a past president of the Coastal Society. Prior to coming to the University of Washington in 1976, Dr. Hershman was associate professor of law and marine studies and coordinator of the Sea Grant Law and Socioeconomics Program at the Louisiana State University. He has law degrees from Temple University Law School.

Michael F. Hirshfield is acting vice president for resource protection programs at the Chesapeake Bay Foundation. He was previously director of the Ecosystem Protection Program at the Center for Marine Conservation (CMC), where he was responsible for developing and overseeing CMC's efforts to ensure integrated management of marine ecosystems, including marine protected areas. Dr. Hirshfield previously was the senior science advisor with the Chesapeake Bay Foundation, where he provided expert scientific and policy advice on the development and implementation of projects related to coastal management, nonpoint source pollution, nutrient and toxics pollution, fisheries management, and estuarine ecology. Before joining the Chesapeake Bay Foundation, he was with the Maryland Department of Natural Resources serving as the director of the Chesapeake Bay Research and Monitoring Division. Dr. Hirshfield received a B.A. in biology from Princeton University and a Ph.D. in zoology from the University of Michigan. From 1981 to 1983, he was director of the Benedict Estuarine Research Laboratory under the Academy of Natural Sciences of Philadelphia.

Eldon Hout is the manager of the State of Oregon Coastal-Ocean Program in the Oregon Department of Land Conservation and Development, where he manages the state's Coastal Zone Management Program. Prior to this, he was the deputy director of the department. Mr. Hout was appointed Oregon's first ocean program manager in 1987 and worked with the Ocean Resources Management Task Force comprised of state agency directors, ocean users, local government officials, and citizen representatives to develop and secure adoption of the nation's first comprehensive plan for the management of ocean resources. Mr. Hout is the state delegate to the Department of Interior Outer Continental Shelf Policy Advisory Committee and the Coastal States Organization (CSO), where he is vice chairman and a member of the CSO Executive Committee. Mr. Hout is a graduate of Stanford University and the Fletcher School of Law and Diplomacy. He taught political science at Pacific University from 1964 to 1974.

Robert L. Howard recently retired from Shell Oil Company after 36 years of service. At Shell he was vice president of domestic operations, exploration, and production and president of Shell Offshore, Inc., an exploration and production subsidiary. He also served as president of Shell Western Exploration and Production, a land-based oil and gas company. Mr. Howard has served on advisory committees to a number of federal agencies. He has a B.S. in mechanical engineering from Rice Institute.

Robert W. Knecht is co-director of the Center for the Study of Marine Policy and a professor at the University of Delaware. He was formerly affiliated with the University of California at Santa Barbara. Prior to his academic career, Mr. Knecht was director of the Office of Ocean Minerals and Energy and assistant administrator for coastal zone management of the National Oceanic and Atmospheric Administration. He served on the Marine Board Committee on Future Uses of the Seabed and has published extensively on ocean policy, coastal zone management, and other topics. He holds a B.S. in physics from Union College and an M.A. in marine affairs from the University of Rhode Island.

Robert C. Repetto is vice president and senior economist at the World Resources Institute (WRI), a nonprofit private organization whose mission is to help governments, international organizations, and private businesses address the question of how societies can meet basic human needs and nurture economic growth without undermining the natural resources and environmental integrity of the biosphere. Prior to joining WRI in 1983, Dr. Repetto was an associate professor of economics and population at the Harvard School of Public Health, where he served since 1974. He is a member of the Science Advisory Board for the Environmental Protection Agency and has served on panels for the National Research Council Transportation Research Board. Dr. Repetto received a B.A. and Ph.D. in economics from Harvard University and an M.Sc. in mathematical economics and econometrics from the London School of Economics. Dr. Repetto's research is in the area of environmental and resource economics.

Alison Rieser is a professor of law at the University of Maine School of Law and director of the Marine Law Institute, where she oversees legal and public policy research on ocean pollution, coastal land use, fisheries, and international maritime relations. She is a consultant to state and federal agencies and editor of the Territorial Sea Journal. Her previous government service includes work with the Environmental Protection Agency and the National Oceanic and Atmospheric Administration. She spent two years at the Woods Hole Oceanographic Institution as a research fellow in marine policy and ocean management. Ms. Rieser has a bachelor's degree from Cornell University and law degrees from George Washington University and Yale Law School.

Katharine F. Wellman is a natural resource economist at the Battelle Memorial Institute in Seattle, Washington, where she specializes in environmental economics as applied to marine resource management and public policy. She has served as a consultant to both federal and state agencies on many issues, including the designation of national marine sanctuaries, wetlands restoration policy, benefits and costs of water quality actions, and salmon restoration and management. Dr. Wellman spent three years at the Woods Hole Oceanographic Institution's Marine Policy Center and two years with the National Oceanic and Atmospheric Administration in Washington, D.C., working on fisheries management, ocean pollution, and marine resource valuation. She has published articles in a number of professional journals and is a lecturer of economics at the University of Washington School of Marine Affairs and Western Washington University. Dr. Wellman has a B.A. and Ph.D. in economics and an M.A. in marine affairs.

George M. Woodwell, is the founder and director of the Woods Hole Research Center, which was organized in 1985. Dr. Woodwell's scientific research focuses on the structure, function, and development of terrestrial and aquatic ecosystems, especially the biotic contributions to the global carbon cycle and effects on climatic change. He is also interested in the application of ecological principles to public affairs and has a strong commitment to the conservation of the marine environment. Dr. Woodwell has a Ph.D. in botany from Duke University. He is a member of the National Academy of Sciences and has served on many National Research Council committees.

APPENDIX
B

Background Paper:
Issues in Marine Area Governance
and Management

PREFACE

In April 1993, the Marine Board of the National Research Council (NRC) sponsored a forum on ocean issues. A broad spectrum of representatives of private industry, public agencies, public interest groups, and the academic ocean policy community were invited to air their views on the need for a national strategy to manage the nation's coastal and ocean resources and space. Based on the proceedings of the forum, the Marine Board requested and received NRC project initiation funds to identify and appraise emerging issues in marine area management. A planning meeting was held in July 1994, which was attended by representatives of interested and active parties in ocean governance and management (see Attachment 1 for list of participants). Based on presentations and discussions at the planning meeting and subsequent comments by the attendees and members of the Committee on Marine Area Governance, the following paper has evolved. It identifies specific issues that need to be addressed in developing the concept and practice of marine area governance and management. This paper is intended to serve as a conceptual framework for an ongoing examination of real-world examples of marine area management projects, with the objective of developing a model for improving ocean governance in the future.

BACKGROUND AND OVERVIEW

The growing national interest in and the appreciation of the ocean and the opportunities to utilize marine resources make it timely to develop a coherent framework to guide the nation's activities in the ocean and coastal regions. New challenges have arisen from changes in national priorities and in the international economic system, including a recognition that good environmental policies make economic sense, the globalization of markets and opportunities, and a new willingness for government to be a catalyst for technological development and economic growth, as well as a steward of the nation's natural resources.

At the same time, demands on the coastal marine environment have been intensifying through the migration of population to the coasts, the growing importance of the coasts and ocean for aesthetic enjoyment, and mounting pressures to develop ocean resources and space for economic benefits (e.g., commercial fisheries, marine aquaculture, marine energy, and mineral resources). All of these factors have created a sense of urgency about developing a coordinated national system for decision-making in the marine arena.

At present, the United States manages its ocean and coastal space and resources primarily on a sector-by-sector basis. For example, to a large degree, one group of laws, agencies, and regulations govern offshore oil and gas; different laws, agencies, and regulations apply to fisheries; still other single-purpose regimes are responsible for water quality, navigation, marine protected areas, endangered species, and marine mammals. It is, perhaps, ironic that while legal regimes for the management of resources operate on a statute-by-statute basis, each area of interest may at the same time be subject to a plethora of other regulatory management regimes. Except for the modest, but important, marine sanctuaries program and a few emerging state programs, the nation does not have the capability to plan and manage ocean regions on an area-wide, multipurpose, or ecological basis; nor is there an agreed upon process for making trade-offs and resolving conflicts among various interests.

Findings from previous Marine Board examinations of issues associated with the nation's ocean space and resources (NRC, 1989, 1990, 1991, 1992; Marine Board 1993) have shown that the absence of a coherent national system of governance for marine resources and uses of ocean space has contributed to economic stagnation and political stalemate in many areas where there are conflicts among competing uses and interests. These studies have also concluded that a more coherent process for governing marine activities and resources would ensure that the nation's ocean ecosystems and living resources were protected and would allow appropriate economic development. In order to achieve these objectives, however, existing and potential conflicts among users of the ocean need to be anticipated and addressed through mechanisms for allocating ocean space and resources fairly and equitably in keeping with national stewardship.

PROJECT DEFINITION AND SCOPE

Marine area governance has two dimensions: a political dimension where ultimate authority and accountability for action resides, both within and among formal and informal mechanisms, i.e., **governance**; and an analytical, active dimension where problem analysis leads to action and implementation, i.e., **management**. The two dimensions need to be integrated according to clear national objectives, which are set forth below. In practice, there is a continuum from the realm of governance to the realm of management. A great many tools are presently deployed in the marine and coastal environment to address one or another aspect of management. But there is no coherent system of governance based on overarching principles.

To fill this need, this forum has attempted to identify principles and goals, as well as the elements of and a process for improving marine area governance. The concepts outlined in this paper are particularly applicable to the marine environment, that is, the zone from high water to the seaward extent of the U.S. Exclusive Economic Zone.[1] The geographic area of concern for this study is the marine environment of the United States, including bays and estuaries, without regard to the jurisdictional authority of the states and the federal government.

The term "marine management area" as used in this paper refers to an area for which coherent plans are developed and measures taken to govern the uses of the area systematically. Marine management areas include sanctuaries, parks, and regional planning and management programs, such as the Gulf of Mexico program. Other attempts by states to plan for the use of their ocean and coastal areas are also under way. Certain statutes, such as the Coastal Zone Management Act, the Fisheries Conservation and Management Act, and the Outer Continental Shelf Lands Act, have also established processes with some of the characteristics of marine area management.

Until recently, prohibiting or severely limiting certain activities, such as exploiting energy resources or commercial and/or recreational fishing, in specific ocean areas has been the primary strategy for controlling development. This strategy does not address the growing problem of protecting the environment and mediating among multiple users in a marine area of intense use for diverse activities, such as transportation, energy and mineral resource development, recreation, commercial fishing, and research. There is a growing recognition of the need for coordinated management of these regions and resources in the best interests of present and future generations.

Defining the basic principles and effective processes for the coordinated governance of ocean and coastal areas is a prerequisite to both sound economic investment and effective environmental stewardship. A coordinated system would make a more reasonable, less adversarial approach to resolving conflicts a realistic possibility. The first step toward developing a new model for coordinated governance is to assess current practices.

NEED FOR IMPROVED GOVERNANCE AND MANAGEMENT PROCESSES

A number of federal agencies now exercise jurisdiction over activities in the ocean and coastal regions. The National Oceanic and Atmospheric Administration

---

[1]The exclusive economic zone is the area 200 nautical miles from each nation's continental boundary; authority over resources in this region is ascribed to the nation.

(NOAA) is responsible for running the marine sanctuary program; for managing fisheries under the National Marine Fisheries Service, which has recently imposed moratoria in areas where resources have been overfished; for implementing the Marine Mammal Protection Act; and for coordinating and overseeing state management of coastal areas through the Coastal Zone Management Act (CZMA). Under the CZMA, NOAA can provide grants to states for coastal management, which have been used in some states for ocean management planning. NOAA also shares responsibility with the Environmental Protection Agency (EPA) for overseeing state pollution control programs for nonpoint source pollution.

The EPA has several ongoing regional planning programs that directly address the uses and management of ocean regions, such as the Gulf of Mexico; they are also responsible for designating and managing ocean dump sites. EPA is the lead federal agency for preparing and promoting implementation of the National Estuary Program.

The Minerals Management Service (MMS) is responsible for the management of Outer Continental Shelf (OCS) energy and mineral resources, which have been subject to legislatively mandated moratoria that place large areas of the U.S. continental margin off limits to exploration for and development of offshore oil and gas.

The National Park Service has a number of responsibilities for ocean resource management, for example, in the Channel Islands, off California, and in the Florida Keys. The U.S. Coast Guard, another interested party is responsible for enforcing laws and regulations in the oceans. The U.S. Department of State's interests in these issues are focused on the international foundations and implications of the regional management of marine resources and uses. States, international agencies, and other countries also have various responsibilities and interests.

The U.S. Department of Defense operates in a number of ocean areas for purposes of carrying out missions related to national defense (e.g., missile ranges, test areas, exclusion areas near gunnery ranges, and areas set aside for maneuvers). The Maritime Administration in the U.S. Department of Transportation's responsibilities are related to ports and marine transportation and safety.

The single-purpose, overlapping, and uncoordinated laws that generally characterize the present system for managing ocean resources is insensitive to the effects of one resource use on other resources and the environment, fail to assess cumulative impacts, and rarely provide a basis for resolving conflicts. In the absence of an overarching governance system, those seeking to utilize ocean resources and space for economic objectives and those concerned with environmental preservation have often reached a stalemate. The societal and economic costs of solving conflicts on a case-by-case basis and the delays inherent in this approach have been high. A brief overview of the most salient problems follows.

Fragmented Government Responsibility

The fragmentation of authority is both horizontal (between agencies at the same level of government) and vertical (between different levels of government). At the present time, management of the marine environment is carried out at local, state, regional, and national (and, in some cases, international) levels of government. In addition, at any given level of government, various functions are carried out through a wide array of agencies and organizations with only limited or sporadic coordination among them, leading to conflicts and inefficiencies.

For example, a port improvement project may require numerous permits from various authorities with inconsistent or even conflicting requirements. Paradoxically, this fragmentation can mean that important issues receive too little attention because they fall through the cracks of various jurisdictions. Thus, a number of agencies may have partial responsibility for managing a marine habitat, but the overall question of habitat protection may never be addressed. Fragmentation also means that real or potential conflicts, either between governmental requirements or proposed uses, are often not anticipated, and when they emerge, effective means of resolving them are not available.

Unresolved Conflicts over Uses of the Marine Environment

Increased competition over uses of the marine environment often leads to conflicts and stalemate. These conflicts have many different manifestations. One obvious conflict is between the exploitation of resources for immediate gains and the less tangible, longer term gains from preserving the environment and protecting ecosystems. The debate over petroleum development in the marine environment and its possible negative impacts on ecological systems, with their varied economic and social values, is an example of a conflict of this kind.

There are also conflicts among competing economic uses as exemplified by the concerns of fishermen about the effects of oil spills on fisheries stocks. Conflicts have also arisen over the economic utilization of common resources, such as the conflict between recreational and commercial fishermen over increasingly scarce resources.

Resolving these conflicts involves making difficult choices among competing needs and interests and raises serious questions about equity, for present and future generations. Public agencies must determine the appropriate balance between immediate economic needs and future needs. Under the present system, decisions are made on a case-by-case basis, often involving costly and lengthy decision processes. Decision makers have no coherent system based on protecting overall national interests to guide them and ensure that the nation's long-term interests are served.

Deterioration of the Marine Environment

The coastal marine environment is under increasing stress. A contributing factor to the imminent deterioration of the ocean and coastal areas is the intensive utilization of space and resources, such as the fisheries exploitation and habitat alteration caused by development in coastal areas. An additional factor is the continuing release of man-made materials into the marine environment, either wastes or the residual from petroleum or pesticides. In addition, there is growing evidence that, in the future, stresses on marine resources from changes in climate will further limit the resiliency of the natural system to survive the pressures of human utilization.

Lost Values and Opportunities

The enormous value of a healthy, diverse, and productive marine environment and its resources to humans is difficult to assess in quantitative terms. The rich biological diversity of the sea, which in some important ways is richer than the diversity of the terrestrial environment, is intrinsically important. For example, of the 32 recognized animal phyla, 15 can be found exclusively in the sea and only one exclusively on land. Also, the basic biological productivity of the rich areas of the sea rivals the productivity of the most fecund tropical jungle.

Beyond this, the marine environment provides a full range of functions and resources. A recent joint publication by a number of national and international environmental groups (Norse, 1993) lists the following important functions of the marine sphere:

- a source of food
- a repository of information for medicinal and related biomedical research
- a source of a variety of raw materials ranging from algae to minerals
- an essential processor of global carbon and other elements
- a venue for aesthetic/recreational activities

The seabed provides many opportunities for resource exploitation, most notably for petroleum, and the surface of the sea handles a major portion of commercial traffic, upon which the global economy rests.

MEETING THE NATIONAL INTEREST

An improved system of marine area governance and management will be effective only if it is perceived as defining and protecting the national interest in the marine environment. The national interest is not synonymous with the federal government's interest. It denotes the fundamental values that the nation as a whole has embraced for the protection and use of the marine environment. The national interest transcends the interests of any single agency mission or special interest group and presupposes a reasonable accommodation among competing interests based on protection of the functioning marine environment.

The complexity and intensity of unresolved conflicts among varying values and economic expectations imposes a number of direct and indirect costs on advancing the national interest. Often the environmental costs are not readily apparent in the short term. For example, the cumulative effects of a series of development activities resulting in substantial habitat alteration may not be obvious until long after the development has taken place.

The national interest in the marine environment can be deduced from a variety of sources and is defined and embodied in national policies for the oceans. Society is made up of groups and individuals with a range of social values and economic expectations. An essential task of marine area governance is to provide mechanisms for identifying and, as much as possible, reconciling these differences.

GOALS AND PRINCIPLES FOR IMPROVED MARINE AREA GOVERNANCE

As human knowledge, values, and needs have changed, so have the demands on institutions of marine governance. The following is a discussion of improved governance of marine areas and how it would benefit society.

Equity

The process for allocating benefits and costs should conform to accepted norms of horizontal and vertical equity. Improved governance would create a level playing field for competing stakeholders and users and would be transparent. Equity would extend to future generations.

Sustainable Development

Some of the greatest failures in marine area management have been failures of sustainability (e.g., depleted fisheries). Sustainable development would provide for the needs of the present without compromising the needs of future generations.

Institutional Effectiveness

Marine governance systems should produce environmentally sound results with the lowest possible expenditure of financial and other resources. Coordinating existing programs to maximize positive synergistic effects and eliminate duplication would greatly increase program effectiveness.

Economic Efficiency

The increased investment in governance would maximize the discounted net economic value of the flow of goods and services produced from resource allocation.

Predictability

A marine governance system should produce expected or predictable results on a timely basis. An improved system would be based on a coherent sense of the factors defining the decision-making process and the nature of any uncertainty regarding results.

Accountability

If authorities and structures for governance and management are clearly demarcated and elucidated, it will be evident who is responsible for particular tasks.

Technologically Achievable Outcomes

Decisions about resource allocation must be supported by existing technology. An improved system of governance would encourage the development and adoption of superior technologies.

Scientific Validity

Governance systems would be supported by known biological, physical, chemical, and ecological facts and principles while recognizing that cultural and social norms might influence decisions.

Terrestrial Connectivity

Marine governance systems would be seamlessly joined to adjacent terrestrial systems, such as coastal zone management programs and state land use regimes.

ELEMENTS RELATED TO IMPROVED MARINE AREA GOVERNANCE

The following elements are necessary for an improved system of governance.

Sense of Place/Ecology

The appropriate criteria for defining the area around which a system of marine governance ought to be organized are difficult to establish. Factors to be considered include size, as well as whether political, economic, or environmental features should determine the boundaries. The following organizing concepts may be useful in approaching the problem:

- **Selectivity**. It is not possible to create a more effective system of governance everywhere at once. Priority should be given to areas where the stress of competing uses is highest or areas with unique ecological value.
- **Ecological Systems**. Issues are often defined and stakeholders engaged on the basis of ecological boundaries. Marine governance and management systems are already emerging around ecological systems, such as the Chesapeake Bay, the Gulf of Mexico, Monterey Bay, the Gulost valuf of Maine.
- **Political Jurisdictions**. Because a marine governance system will have to accommodate a variety of existing institutions, both horizontal and vertical, boundaries by cannot be defined by jurisdictions. The transboundary nature of marine concerns and issues can best be addressed by a system that transcends political boundaries.

Democracy/Public Participation/Transparency

Marine resources have been exploited by humans since earliest times. In this century, attempts have been growing to manage the use of specific resources, such as fisheries and oil. No system of governance, however, has evolved to ensure the participation of all stakeholders (analogous to governance systems in the terrestrial environment). An essential feature of marine area governance and management must be the "ground-up" participation of all stakeholders in the governance process.

Multiple Issues

Very few marine issues can be addressed in isolation. Actions that affect one resource necessarily affects other resources. Wise management of a particular resource will require that action be taken in a variety of arenas. The current era of single resource management must give way to an era that allows—and even forces—management schemes that take into account the full spectrum of economic, social, and ecological uses of the marine environment.

Institutional Connectivity/Structure

Just as the ground-up involvement of stakeholders is only beginning to emerge in the marine context, so is the development of robust structures for governance mechanisms for ensuring their interconnected operation. The current system is characterized by government programs that administer single resource management at the local, state, or federal level. To expect that these separate programs will disappear and be replaced by a superagency of government endowed with all responsibility and authority in unrealistic, perhaps even undesirable. However, one can imagine substantial improvement by establishing strong links in analysis and decision making from one program to another. In essence, improved connectivity or coordination will help develop a strong governance process for the marine environment.

Integrated Decision Making

In a coordinated institutional setting, where synergies are created through mutually reinforcing decision making across a range of programs, single issue decision making would no longer be feasible. Institutional structures that addressed the full range of issues would require partnerships among the affected agencies.

Conflict Resolution

Even if decision making is integrated, frequent conflicts will still arise. A major objective of the improved arrangement is to resolve conflicts expeditiously. The existence of a coherent and transparent process will, at least, make outcomes somewhat predictable. This predictability in itself, should reduce delays and uncertainties associated with protracted disputes.

Adaptive Management

Few management systems are once-through exercises. Human values change; scientific understanding increases; threats ebb and flow; and human needs evolve. Improved governance must incorporate adaptive environmental management processes in order to be responsive to constant changes and to make better use of scientific information to protect the environment. Outcome-based monitoring can be an important tool for adaptive management.

PROCESS FOR IMPROVED MARINE
AREA GOVERNANCE

Improvements in marine area governance and management will not necessarily entail the creation of new institutions to replace or supplement existing ones. Rather it will be a process that brings together and harmonizes existing programs. The process will be based on the concepts of adaptive environmental management, in general, and integrated coastal management, in particular.

Elements of the Implementation Process

Critical elements of the process include the following:

- There must be a clear statement of goals. Bringing different entities to-gether for cooperative management requires a clear understanding of the nature of the problem(s), identification of probable causes, and clear and unambiguous goals.
- The geographic (or ecological) management area needs to be carefully delineated. See discussion under "Sense of Place/Ecology" above.
- Mechanisms for involving all relevant stakeholders in the governance pro-cess need to be designed. Stakeholders include government, public inter-est organizations, and private parties. The experiences of the National Estuary Program may provide useful models for involving stakeholders.
- In most situations, the process should be a joint state-federal effort. In virtually every setting, the state and local interests are as substantial as those of the federal government. In several regions, the states have al-ready exercised leadership in ocean planning and governance. Integrating the management functions of a range of government organizations will be critical.
- Systems of marine governance should be designed to foster innovative responses to management needs and opportunities for resource utiliza-tion. Robust programs are part of a coherent system are more open to innovation than single-purpose or fragmented programs.
- Processes that facilitate the incorporation of scientific information into all aspects of decision making should be established. Good decisions in this sphere, as in many others, need to be based on the most reliable, scientific research. Recognition of the importance of scientific information must be built into the governance process.
- Success should be clearly monitored and evaluated. Careful monitoring can also contribute to scientific knowledge. Any process of marine area governance must incorporate monitoring and evaluation systems to assess and report on the state of the environment or targeted resources. These systems will need to be based on standards and parameters that can mea-sure success or failure in reaching agreed upon goals.

Building on Previous Work

The processes of environmental management have been well described by others. This study need not dwell on them or repeat what has already been done. A final report might include a chapter or appendix briefly summarizing and re-viewing the relevant literature and describing the general precepts of adaptive environmental management to governance in the marine area.

EXISTING APPROACHES TO MARINE MANAGEMENT

Existing marine management areas include a broad spectrum of coastal and marine areas with attempts at managing resources for sustainable use, safeguarding ecosystem health and biodiversity, and/or providing a framework for the utilization of resources and space with a minimum of conflict. Marine management areas are not merely marine parks or sanctuaries. They range from small closed areas or harvest refugia, designated to protect specific resources or habitat types or to prohibit specific resource development activities, to extensive coastal and/or marine areas that integrate the management of many species, habitats, and uses in a single plan. The major categories of marine management areas are described in Attachment 2. These categories are clearly not mutually exclusive; many can be and often are used in conjunction with each other.

Alternative Approaches to Marine Management Areas

A fully developed system that meets all of the objectives and contains all of the elements discussed above will necessarily evolve over time in response to actual experience. However, there are opportunities for moving forward now based on the experiences of existing marine management areas. This section sets forth changes that are presently under way and can be used as starting points for developing a model for marine area management.

- Special Area Management Program under the CZMA. The Coastal Zone Management Act (CZMA) of 1972 (as amended) includes a provision allowing for the designation and establishment of special area management programs by the coastal states as part of their coastal zone management (CZM) programs. Relatively minor changes in the legislative language could allow coastal states to undertake initial efforts to implement the governance and management systems described in this paper. Other aspects of the CZMA offer vehicles for ensuring coordination and integration of decision making processes and need to be examined as models for new ocean governance regimes or approaches.

- Marine Sanctuaries. The National Marine Sanctuary Program is a unique federal that tool offers many opportunities for improved governance. Although the primary objective of the sanctuaries program is to protect specific marine resources of exceptional value, it does this through a process that allows for the analysis and management of a multiplicity of uses within the identified area of the marine sanctuary. Evaluating a sanctuary program in relation to the objectives and elements identified in this issues paper would be extremely valuable. The question of whether certain management actions within sanctuaries create greater opportunities for improved governance outside of the sanctuary boundaries is of particular interest.

- Moratoria for Oil and Gas Leasing and Fishing. Considerable controversy has surrounded the leasing of outer continental shelf (OCS) areas for oil and gas development and limitations on fishing, such as the limitations imposed on striped bass fishing in Maryland in 1995. Moratoria are one strategy for marine management that prohibits certain activities. Benefits have accrued to the environment and/or to certain species from these actions. It would be useful to investigate the benefits and costs associated with using moratoria as tools for managing selected marine activities.
- New Approaches to Management of Marine Areas: the National Estuary Program and State Ocean Plans. At least two new approaches to the management of marine areas are at an early stage of development. These are the National Estuary Program and several formalized state ocean planning programs. Both programs were created primarily in response to initiatives at the state level. Both operate in clearly defined geographical areas. Both have thus far been developing planning processes to deal with a wide range of issues.
- International Institutional Structures. New institutional structures at the international level, as exemplified by the Law of the Sea and the Gulf of Maine initiative, illustrate that the broad outlines of a new approach to marine issues are beginning to emerge. The Convention on the Law of the Sea III came into effect in 1994. This convention assigns responsibility to each nation for exercising governance functions over resources in its Exclusive Economic Zone (EEZ). These examples illustrate that government is already trying to improve governance and that there are opportunities to build on these beginnings.

IMPLEMENTING AND MEASURING SUCCESS

Improved marine governance and management need not involve the creation of new superagencies to assume the responsibilities currently dispersed among a wide range of agencies. However, some mechanism must evolve to link the elements of improved governance that this paper identifies and to oversee their implementation. This may require the development of new institutions (or responsible agents within existing institutions) to carry out relatively limited tasks of coordination or to assign tasks to existing agencies. The following are some of the tasks that may need to be carried out by a single identified agency, at either the federal or state level.

Planning Coordination

Although detailed, specific planning and analysis might be done by a range of organizations, a commitment must be made to integrating and harmonizing these efforts across the multiplicity of agency functions and shareholder conflicts. The goal is to identify and implement effective mechanisms for integrating decision making in order to minimize and, wherever possible, resolve conflicts, and to maximize cooperation and sharing of resources to avoid inefficiencies and delays.

Budget Coordination

Agencies may retain budget autonomy in a traditional sense, but there must be some oversight of priorities in the allocation of funds to ensure that the objectives of the integrated planning process are met.

Monitoring and Evaluation Management

A crucial instrument for improving management is monitoring performance in the field to track whether the objectives and goals of the system have been achieved and to make corrections in the original course, as appropriate and necessary. Criteria should provide appropriate measures for this purpose.

Public Accountability

Both the individual agencies participating in an integrated management process and any new coordinating mechanism should be publicly accountable for the results (or lack thereof) in terms of specific objectives and goals.

CONCLUSIONS

Findings

More and more marine management areas (e.g., marine sanctuaries and marine and coastal parks) are being designated. Other attempts to manage multiple-use marine areas are also under way (e.g., designated national estuaries under EPA's National Estuary Program, areas designated for special management in state coastal plans). However, no overarching national policy has been articulated to guide the long-term marine management of these areas. There is a critical need for guiding principles for governance and management.

Although moratoria can be appropriate and effective responses to specific issues, a more inclusive approach may be needed to manage a complex range of activities in these areas in the best interests of the nation—present and future. An inclusive approach must be guided by principles and policies that reflect the long-term national interests in ocean and coastal regions and resources.

Plan of Action

The Marine Board of the National Research Council will assess and develop guiding principles for the governance and management of marine areas. The project will be carried out by a committee of experts that will undertake case studies of representative examples of marine management areas and, based on the findings from the case studies, will develop guidelines for improving governance and management of marine areas to both in terms of environmental stewardship and the development of ocean resources. The committee will prepare a published report that will propose models and methods for marine area governance and management to guide federal and state agencies with jurisdiction over ocean areas and uses.

The case studies will be chosen to represent diverse management areas and geographic regions. Case studies should include a marine sanctuary where eco-system management concerns conflict with recreational use or resource development (e.g., the Florida Keys); an area where a moratorium on energy resource development is in effect to assess the benefits and costs of the moratorium and explore other options for resolving disputes over resource development (e.g., Gulf of Maine); and an ocean area of intense use for various activities, including commercial marine transportation or commercial fisheries, to learn more about the relationship between marine traffic management and marine environmental management (e.g. Southern California).

Each case study will be assessed with regard to (1) the effects of local, state, and federal regulations; (2) ecological and biological issues; (3) the potential for commercial or recreational uses; and (4) the social, cultural, and economic context. Criteria to guide the conduct of case studies of marine management areas are based on the analysis in this issues paper and the deliberations of the committee. Regional perspectives and expertise will be sought through meetings held in the case study areas. Federal agencies with responsibilities for marine management will be asked to designate liaisons to the committee to provide an avenue for the exchange of information and also to lend their expertise.

Following the case studies, the committee will distill lessons learned from the case studies and other activities and develop conclusions and recommendations. Based on these findings, the committee will prepare a published report proposing models for marine area governance and management to guide federal and state agencies with jurisdiction over ocean areas and uses.

REFERENCES

Marine Board. 1993. Marine Board Forum: The Future of the U.S. Exclusive Economic Zone. Marine Board, National Academy of Sciences. Unpublished proceedings. April 10, 1993.

National Research Council (NRC). 1989. Our Seabed Frontier. Washington, D.C.: National Academy Press.

NRC. 1990. Interim Report of the Committee on Exclusive Economic Zone Information Needs: Coastal States and Territories. Washington, D.C.: National Academy Press.

NRC. 1991. Interim Report of the Committee on Exclusive Economic Zone Information Needs: Seabed Information Needs of Offshore Industries. Washington, D.C.: National Academy Press.

NRC. 1992. Working Together in the EEZ: Final Report of the Committee on Exclusive Economic Zone Information Needs. Washington, D.C.: National Academy Press.

--------------------------- **ATTACHMENT 1** ---------------------------

MARINE MANAGEMENT AREA GOVERNANCE

Planning Meeting
July 28–29, 1994
Washington, D.C.

William M. Eichbaum, Co-Leader
World Wildlife Fund

Robert W. Knecht, Co-Leader
College of Marine Studies
University of Delaware

Tundi Agardy
World Wildlife Fund

Anne Aylward
National Commission on
Intermodal Transportation

Robert Bendick
New York Department of
Environmental Conservation

Jeffrey R. Benoit
Office of Ocean and Coastal Resource
    Management
National Oceanic and Atmospheric
    Administration (NOAA)

Brock B. Bernstein
EcoAnalysis

Darrell D. Brown
Environmental Protection Agency

Richard H. (Rick) Burroughs
Department of Marine Affairs
University of Rhode Island

Ralph Cantral
Department of Community Affairs
Florida Coastal Management Program

Francesca Cava
Sanctuaries and Reserve Division,
    NOAA

Sarah Chasis
Natural Resources Defense Council

Biliana Cicin-Sain
College of Marine Studies
University of Delaware

Jacob J. Dykstra
Point Judith Fisherman's Cooperative

Charles N. (Bud) Ehler
Office of Ocean Resources Conservation
    and Assessment, NOAA

Tim Eichenberg
Center for Marine Conservation

Richard A. Fitch
Offshore Operators Committee
New Orleans, Louisiana

Tom Fry
Minerals Management Service

Eldon Hout
Oregon Ocean Program

Marc J. Hershman
School of Marine Affairs
University of Washington

Lee Kimball
Washington, D.C.

Cynthia Quarterman
Minerals Management Service

Paul Stang
Minerals Management Service

James P. Ray
Environmental Sciences Division
Shell Oil Company

Peter Shelley
Conservation Law Foundation of
    New England

Stephanie R. Thornton
Coastal Resources Center
San Francisco, California

--------------------------- **ATTACHMENT 2** ---------------------------

MARINE MANAGEMENT AREAS

Prepared by Tundi Agardy
World Wildlife Fund

Marine management areas constitute a broad spectrum of coastal and marine areas that are afforded some level of protection for the purpose of managing resources for sustainable use, safeguarding ecosystem function and biodiversity, and/or providing a framework for the use of resources and space with a minimum of conflict. Marine management areas are not merely marine parks or sanctuaries—they range from small closed areas or harvest refugia designated to protect a specific resource or habitat type to extensive coastal zone areas that integrate the management of many species, habitats, and uses in a single, all-encompassing plan. Seven major categories of marine management areas are described below. These categories are clearly not mutually exclusive; they can be and often are used in conjunction with one another.

Category 1: Closed Areas

Closed areas include refugia where harvesting fish or other living marine resources is prohibited; moratoria, areas closed to resource exploration, development, or harvesting; and areas where a certain class of use is restricted for the purpose of ensuring the sustainability of resources or in response to concerns about potential environmental damage. Closed areas differ from sensitive sea areas (Category 4) in that management of a specific type of use is the main objective for establishing a moratorium or designating a site as closed. Closed areas can be and often are temporary or seasonal.

Objective: To allow for replenishment of stocks of renewable resources, such as fish and shellfish, by prohibiting harvest at sites critical to the target species, or, in the case of nonrenewable resources, to protect sites by prohibiting mining and exploration.

Criteria: Closed areas are areas where if specific activities are restricted specifically for the purpose of protecting a stock or population of one or more species or protecting a particular habitat from possible damage.

Examples: Fisheries harvest refugia designated off the coast of California; outer continental shelf (OCS) moratorium areas in U.S. continental shelf waters where oil and gas exploration and development activities have been suspended; "no-take" zones in New Zealand waters.

Category 2: Research and Monitoring Areas

Research and monitoring areas are either experimental controls or sites for environmental monitoring or are protected as *in situ* natural laboratories to support basic research in ecology, fisheries, oceanography, etc. Research sites are managed specifically for the purpose of controlling research variables or allowing for intersite comparisons and can either be independent entities (e.g., long-term ecological research sites) or core areas within multiple use reserves.

Objective: To provide protected areas where certain anthropogenic impacts can be controlled for the purpose of experimental or environmental research.

Criteria: A marine managed area is considered a research area if it is managed specifically for the purpose of protecting the site so that research can be undertaken with a minimum of extrinsic variability.

Examples: Long-term ecological research sites, National Estuarine Research Reserves, core areas within biosphere reserves, scientific research zones in multiple use marine parks (e.g., Great Barrier Reef Marine Park).

Category 3: Marine Sanctuaries and Marine Parks

Marine protected areas, such as marine sanctuaries and traditional marine parks, constitute a broad and complex assemblage of marine management areas. The World Conservation Union formally recognizes 10 classes of marine protected areas, including: strict nature reserves, national parks, natural monuments, wildlife sanctuaries, protected seascapes, resource reserves, natural biotic areas or anthropological reserves, multiple use management areas, biosphere reserves, and world heritage sites. Clearly some of these categories overlap with the seven main categories described here. Nonetheless, the feature common to all marine parks and sanctuaries is that they are established to accommodate particular uses while conserving the coastal or marine ecosystem and its processes. Marine parks and sanctuaries range from seaward extensions of coastal terrestrial parks, to ecosystem-based multiple use marine parks, sanctuaries, and biosphere reserves.

Objective: To protect coastal and marine habitats, conserve ecosystem processes, and allow for the sustainable use of marine resources and space with a minimum of conflict, often for the primary purpose of increasing or maintaining recreational and aesthetic value.

Criteria: A marine management area falls under the heading of marine park, reserve, or sanctuary if it is officially designated as such by local or national authorities.

Examples: National Marine Sanctuaries (e.g., Florida Keys, Stellwagen Bank, U.S.S. Monitor, Gulf of the Farallones, Cordell Bank, Channel Islands, Flower

Garden Banks); Great Barrier Reef Marine Park (Australia); Mafia Island Marine Park (Tanzania); El Nido Marine Park (Philippines).

Category 4: Sensitive Sea Areas

The International Maritime Organization (IMO) recognizes sensitive sea areas as areas that need special protection through action by the IMO because of their ecological or socioeconomic significance and their vulnerability to damage by maritime activities. Sensitive sea areas include coral reef areas or temperate sounds where ship transit is prohibited for reasons of safety and environmental sensitivity.

Objective: To safeguard particularly vulnerable habitat types, such as diverse coral reef systems, by declaring areas off-limits for certain types of shipping and boating and resource extraction.

Criteria: A marine management area is considered a sensitive sea area if it is of high biological value, vulnerable, threatened, and officially designated as such by the IMO.

Examples: Portions of the Great Barrier Reef (Australia); eastern Arabian Sea; Bay of Bengal.

Category 5: Regional Seas and Large Marine Ecosystem Areas

Regional seas are formally recognized by the United Nations Environment Programme as enclosed or semi-enclosed seas that fall under the jurisdiction of more than one nation. Regional seas become marine managed areas when bilateral or multilateral agreements are drawn up to control pollution, develop cooperatively protected areas (e.g., transboundary reserves), and allow for joint management of endangered species or commercially important renewable resources. Large marine ecosystems are areas that represent a coherent ecological unit (whether enclosed or semi-enclosed seas or biogeographically distinct oceanic systems) that sometimes form the basis for regional seas agreements.

Objective: To provide a framework for cooperative management of resources, multilateral and transboundary protected areas, and/or joint pollution control in marine areas bounded by more than one coastal nation.

Criteria: A marine management area is deemed a regional sea if an international instrument is developed to codify willingness for joint conservation or management of a semi-enclosed or enclosed marine area.

Examples: Mediterranean Basin (Barcelona Convention); Caribbean Sea (Cartagena Convention).

Category 6: Integrated Management Zones

Integrated management zones include state-administered coastal zone planning areas and exclusive economic zones managed by federal authorities. Management of state or provincial coastal zone areas tends to be coordinated and integrated because these areas usually fall under the purview of a single management authority in each state, whereas federally-managed exclusive economic zones may be administered by many different agencies.

Objective: To coordinate management of ocean space, coastal land use, resource extraction, and other activities that take place in or impact a coastal zone for the purpose of minimizing conflict, maximizing management efficiency, and safeguarding the resource base and ecological processes.

Criteria: A marine management area is considered an integrated management zone if specific legislation and administrative structures exist to coordinate all conservation and resource use activities in that area. Successful integrated management zones will be those for which conservation and management plans are drawn up and implemented by all shareholders willing and able to join in the process.

Examples: state coastal management areas; formal national coastal and ocean plans (outside the United States), exclusive economic zones.

Category 7: High Seas under the Law of the Sea Treaty

Although the high seas technically constitute a global commons and are therefore not a managed marine area, international treaties and codified customary law create a cooperative management regime for states that sign and ratify these agreements.

APPENDIX
C

Participants in Committee Meetings

WASHINGTON, D.C.
August 7–9, 1995

Guests

Stephanie Campbell, National Ocean Service, National Oceanic and Atmospheric Administration

Richard F. Delaney, Urban Harbors Institute

Patty Dornbusch, National Ocean Service, National Oceanic and Atmospheric Administration

Porter Hoagland, Woods Hole Oceanographic Institution

DeWitt John, National Academy of Public Administration

Evie Kalketenidou, Maritime Administration

Thomas R. Kitsos, Minerals Management Service

Cynthia Quarterman, Minerals Management Service

Andrew Solow, Woods Hole Oceanographic Institution

Harold M. (Hal) Stanford, National Oceanic and Atmospheric Administration

Stephanie Thornton, Coastal Resources Center

W. Stanley Wilson, National Ocean Service, National Oceanic and Atmospheric Administration

IRVINE, CALIFORNIA
November 29–December 1, 1995

Contractors

Porter Hoagland, Woods Hole Oceanographic Institution

DeWitt John, National Academy of Public Administration

Richard Minard, National Academy of Public Administration

157

Presenters

OCS Oil and Gas Leasing Activities

Ellen Aronson, Minerals Management
    Service

State Ocean Planning

Brian Baird, California Resources Agency
Craig MacDonald, State of Hawaii

Panel of Users

Jerry A. Aspland, AMOCO (retired)
    [*marine transportation*]
John Dorsey, City of Los Angeles
    [*municipal waste*]
Robert Fletcher, Sportfishing Association
    of California [*recreational fishing*]

Mark Gold, Heal the Bay [*recreation*]
Robert Kanter, Port of Long Beach
    [*port issues*]
John Patton, Santa Barbara County
    [*county view of offshore resources
    development*]
David Ptak, Pacific States Marine
    Fisheries Commission/Chesapeake
    Fish Company [*commercial fishing*]

The Channel Island Experience

John Miller, director, Channel Islands
    National Marine Sanctuary

Other Guests

Stephanie Thornton, Marine Board,
    National Research Council

MIAMI, FLORIDA
February 12–14, 1996

Contractors

Porter Hoagland, Woods Hole
    Oceanographic Institution
DeWitt John, National Academy of Public
    Administration
Richard Minard, National Academy of
    Public Administration

Guests

Panel of Government Officials

James Bohnszak, National Marine
    Fisheries Service
Billy Causey, Florida Keys National
    Marine Sanctuary
George Garrett, Monroe County Marine
    Resources
Fred McManus, Environmental Protection
    Agency

G.P. Schmahl, Florida Department of
    Environmental Protection

Panel of Local Leaders and Experts

Mike Collins, fishing guide (chair, Florida
    Keys National Marine Sanctuary
    Advisory Council)
J. Allison DeFoor, lawyer (member,
    Florida Keys National Marine
    Sanctuary Advisory Council)
Debbie Harrison, World Wildlife Fund
Tony Iarocci, commercial fisherman
John Ogden, Florida Institute of
    Oceanography

Other Guests

Wesley Marquardt, U.S. Coast Guard
David Suman, University of Miami

BOSTON, MASSACHUSETTS
April 17–19, 1996

Contractors

Porter Hoagland, Woods Hole Oceanographic Institution

DeWitt John, National Academy of Public Administration

Hauke Kite-Powell, Woods Hole Oceanographic Institution

Andrew Solow, Woods Hole Oceanographic Institution

Presenters

Living Marine Resources (Non-Groundfish) Panel

Cliff Goudey, MIT Sea Grant [*ocean mariculture*]

Larry Hildebrand, Environment Canada [*Sable Island*]

John Williamson, New Hampshire Commercial Fisherman's Association [*harbor porpoise issues*]

James Wilson, University of Maine [*Maine lobster*]

Marine Ecosystem Governance Panel

Brad Barr, Stellwagen Bank National Marine Sanctuary

Christine Gault, Waquoit National Estuarine Research Reserve

David Keeley, Maine State Planning Office/Gulf of Maine Council

Judith Pederson, Massachusetts Institute of Technology, Sea Grant College Program

Robert Wall, Regional Marine Research Program for the Gulf of Maine

Living Marine Resources (Groundfish) Panel

Kathy Holmstead, Holmstead Marine Enterprises [*offshore gillnet fishing*]

Peter Partington, Department of Fisheries and Oceans, Canada [*Nova Scotia fisheries*]

Andy Rosenberg, National Marine Fisheries Service, New England Region

Peter Shelley, Conservation Law Foundation

Guests

Stephanie Thornton, Marine Board, National Research Council

OAKLAND, CALIFORNIA
June 9–11, 1996

Presenters

San Francisco Bay Demonstration Project Panel

David Adams, chief wharfinger, Port of Oakland

CAPT Thomas Richards, San Francisco Bay Demonstration Project

Will Travis, San Francisco Bay Conservation and Development Commission

CAPT Carl Bowler, San Francisco Bar Pilots Association

Seport Plan Concept and Port Dredging Issues Panel

Jim McGrath, Port of Oakland

Kay Miller, Alameda Reuse and Redevelopment Authority

Brian Ross, Environmental Protection Agency

Will Travis, San Francisco Bay Conservation and Development Commission

Guests

Brian Baird, California Resources Agency
Marcia Brockbank, San Francisco Estuary
    Project

Patty Dornbusch, National Oceanic and
    Atmospheric Administration
David Evans, National Oceanic and
    Atmospheric Administration

APPENDIX
D

Financing Options

BONDS

Bond financing allows private or public bodies to spread the burden of cost for a program or project over a long period of time. In some instances, the program created as the result of a bond issue provides enough revenue to pay off the bond; in other cases, general tax revenues are required for repayment. Thus, a project must have enough political support within a jurisdiction to win approval because the people of the designated area pay for the bond either through higher taxes or user fees. States and "special districts" or regions are authorized to issue bonds yielding interest that is exempt from federal taxation to finance programs with some recognized national public interest.

TAXES

Governments raise revenue mostly through taxes. Taxes can be grouped into three general categories: income taxes, property taxes, and taxes on goods and services. Personal and business income is taxed by national, state, and some local governments. Rates may be progressive or flat. Taxing income creates incentives for taxpayers to reduce their liabilities by changing the form or the place income is earned or by changing the amount of effort they expend to earn taxable income.

Taxing personal property, such as real estate, boats, and automobiles, is common among local governments. Exemptions for government property and lease-holds are often granted, and some personal exemptions may be granted as well. Real estate taxes are thought to be appropriate for financing local government services enjoyed by all residents (police and fire protection, for example) but are increasingly being supplemented by specific user fees.

Taxes on goods and services are charged by most states and include taxes on many consumer items. Broad-based sales and value-added taxes are like consumption taxes or income taxes with exemptions for savings. In addition, excise taxes are sometimes levied on specific commodities to provide revenues for government programs related to the commodity (federal gasoline taxes) or to discourage consumption (taxes on alcohol and tobacco, for example). If the market price of an activity does not fully reflect its full economic costs, taxing it may improve the allocation of resources by reducing excessive demand.

Examples of Taxes

Motor Fuels and Petroleum Production Taxes

Because recreational boating has an impact on water quality, marine fuel taxes may be viewed as an equitable method for financing both the capital and operating expenses of water quality improvements. This tax applies to both resident boaters and people who use their boats for transportation, thereby making the tax more equitable and harder to circumvent than boat registration fees. Commercial carriers are currently assessed a federal marine fuels tax that is used to finance the Inland Waterways Trust Fund. State taxes on marine fuels can be assessed on both commercial and recreational users.

Tourist Development and Impact Taxes

The quality of marine resources in the United States is affected by activities that support seasonal tourism. For example, the use of package plants by hotels, motels, and restaurants has been suspected of increasing nutrient loadings. Revenue generated by taxes on lodging and meals can be used to offset some of the costs of tourist-related impacts.

Foodfish and Shellfish Taxes

Another method of financing activities to protect water quality and enhance foodfish and shellfish resources is a foodfish and shellfish tax. Washington state levies a tax on the person with first possession of foodfish or shellfish for commercial purposes after it has been caught. The state is currently investigating removing the exemption for aquaculture. The tax rate is on a variable scale by type of fish or shellfish.

In Maryland and Georgia, a shellfish tax is levied on the leasing of commercial shellfish harvesting areas. In Virginia, a saltwater take fee is applied to "taking" or harvesting oysters by the commercial shellfish industry. Virginia requires both vendors and fishermen to document the amount of oysters sold.

Proceeds from these taxes are generally used for resource management and assisting commercial fishermen. Activities to improve water quality could also be funded as resource management and enhancement. Although the burden of these taxes in Maryland and Georgia falls directly on commercial fishermen, it is assumed that some of the burden of the tax is shifted to consumers in the form of higher shellfish prices. Washington's tax is imposed directly on consumers.

Aquatic Lands Leasehold Tax

In Washington state, a leasehold tax on all public lands leased to private parties (including aquatic lands) is levied at both the state and local levels. Properties are charged at a rate of 12.84 percent on the contract or true rental value of lands that are exempt from property taxes.

Pollutants Tax

This category covers a range of taxes charged on specific pollutants. The state of Florida has, for example, three pollutant taxes that are allocated to various water quality related trust funds. A coastal protection tax of two cents per barrel is charged for pollutants produced in, or imported into, the state. Under this tax, pollutants include petroleum products, pesticides, chlorine, and ammonia. Proceeds from this tax are allocated to the Coastal Protection Trust Fund to be used by the Florida Department of Environmental Resources for cleaning up spills. The tax will remain in force until the balance of the trust fund reaches or exceeds $50 million. If the U.S. Department of the Interior approves offshore oil drilling in the waters off the Florida coast, the cap on the Coastal Protection Trust Fund will be increased to $100 million.

Nine other states also impose some type of pollutants tax. Washington state has a pesticide tax of 0.7 percent of the wholesale value of the product. Consideration has been given to basing the amount of the tax on toxicity, persistence and bioaccumulation of the pesticide. Activities toward which the proceeds from this tax could be directed include pollution control, household hazardous waste programs, wetlands, storm water, environmental education, and environmental enforcement.

Impact Taxes

Several states currently charge taxes on goods or activities that have a perceived impact on public resources. Rather than having a number of specific fees, an impact tax could cover all public costs associated with development, for example. Alternatives for levying impact taxes on development are: per unit charges for new construction (i.e., per living unit, square foot, or land unit area), an excise tax on construction materials, a gross receipts tax on contractors and developers, and a rezoning tax based on the category to which the land is zoned and the number of acres. The amount of revenue generated by impact taxes would depend on the amount of development. The funds could be used to finance water quality and habitat enhancement related activities, such as wastewater and storm water treatment and wetlands preservation or mitigation.

Surtaxes on Sales Taxes

General sales taxes may provide substantial revenue that is allocated via various appropriation and revenue sharing programs. These revenues may not always provide for the complete financing needs of a state or region, and in some cases governments may authorize additional sales taxes to meet specific needs. This is the case in Florida, where in addition to general sales taxes the state of Florida allows certain "discretionary sales surtaxes" to be levied.

Impact of Taxes

Taxes, whether on income, pollutants, resource uses, or sales of goods and services, can provide a significant source of revenue for marine resource management programs. Unfortunately, some taxes, such as personal income taxes, tend to be allocated to general revenue funds from which monies must be appropriated for specific programs. There is no guarantee that funds allocated one year will be available in subsequent years, making it difficult to finance long, ongoing programs with these taxes. Sales and use taxes can be more easily tied to specific programs, as has been done in Florida. These taxes can also be used to discourage the use of, or to mitigate the impacts of, goods and services that have an adverse effect on the environment.

The major disadvantages of taxes are their unpopularity and their often unequal impact on various segments of the population. Generally, taxes require voter approval, and gaining public acceptance can be a costly process. Sales and use taxes have been criticized as regressive because all goods and services are taxed at the same rate regardless of the purchaser's ability to pay. In addition, although a few sales and use taxes can have a relatively small impact, in combination they can create a significant burden. Implementation of numerous taxes can also create additional administrative burdens and costs.

Whether a tax is equitable depends both on the item on which the tax is assessed (i.e., property, goods, or services) and the way the tax is implemented. Most of the sales and use taxes identified in this section have been considered equitable taxes for the purposes of natural resource quality enhancement and environmental protection on the basis that the goods and services on which the taxes are levied affect water quality and the environment. These taxes are merely a variation of the "polluter pays" principle.

GRANTS AND LOANS

Marine resource management programs should be viewed as unique cooperative arrangements between the federal government and states or regions. A number of sources of state and federal funds are available for financing resource management, protection, and restoration. Information on funding sources is available in the Catalog of Federal Domestic Assistance published by the U.S. Office of Management and Budget (OMB, 1994).

INNOVATIVE SOURCES OF FUNDING

Financing alternative marine area governance and management programs will require the creative use of financial resources. Financing alternative marine area governance programs solely through federal and state taxes, grants, low interest loans and cost-sharing programs, and bond issues is becoming increasingly difficult. As pressures on government budgets increase and many funding sources are reduced or eliminated, alternative sources of financing will have to be developed.

Alternative financing is not complicated, but it has been shrouded in mystery for many years because, as long as federal and state funding sources could be relied upon, creative financing was not necessary. One basic premise of finance is identifying a steady, reliable source of revenues to repay the costs of implementing a project.

Revenues are streams of funds collected periodically, but reliably, for services or benefits rendered. Revenues can be generated in many ways, for example, user fees, impact fees, special surcharges, and utility rates. Revenue streams are ideally suited to support the ongoing operations and management requirements of a management program. Once a revenue stream has been dedicated to pay for the operations and management and debt repayment requirements of a management program, then sources of capital can be identified and committed to the program.

Capital is usually a lump sum of funds used to build a facility or other capital asset. Most capital (or commitment to provide capital) arrives at the beginning of a management program and is used to develop program infrastructure. Sources of capital for a management program include the bond market or any capital market; banks and other financial institutions, such as insurance, finance, and leasing companies; and private investors, such as corporations, foundations, and individuals. Capital will not be invested in a program, however, until a steady, reliable source of revenue can be identified and dedicated to the program for debt repayment and maintenance.

Just as a diverse group of people will enjoy the opportunities provided by marine area management programs, so too should diverse sources of funding be used to pay for these programs. No single source of funds should be relied on. A few innovative ideas for identifying steady, reliable sources of revenues and capital to support management programs are outlined below.

Establish special assessment districts (e.g., watershed or ecosystem management districts). A special assessment district is an independent government entity formed to finance governmental services for a specific geographic area. These districts can range in size from a city block to a multijurisdictional area. Special districts focus the costs of enhanced services on the beneficiaries of those services by separating benefitted taxpayers from general taxpayers. Residents of special districts pay taxes (usually in the form of increased tax rates) to finance improvements from which they will benefit. If, for example, citizens in a certain geographic area are interested in reclaiming area wetlands or enhancing recreational opportunities by improving the quality of a waterway, a special district can provide needed structure, management, and financing.

Special districts have the power to levy taxes and to collect fees and special assessments to pay for the development and operation of management programs. Special districts may issue revenue bonds to finance revenue-generating programs, such as fee-based wetland preserves or fee-based fisheries management. Special districts can issue debt, independent of region or state, thus reducing the burden on general debt capacity.

Dedicate a sales tax surcharge on certain products, such as prepared foods and beverages, to management programs. A surcharge is added to the existing prepared food and beverage sales tax. Revenues generated are dedicated to specific beneficial use projects. The surcharge may be time-limited (e.g., 10 years), with optional renewal by the legislature.

Price at full cost the public sector service fees associated with coastal and marine resource management programs (e.g., commercial fisheries management). Existing fee systems associated with public sector oversight programs are modified to cover most or all of the costs of a program. The fee system should ensure that staff, supplies, and overhead costs associated with program development and implementation are covered.

Implement tax-increment financing (similar to a special assessment district). This technique requires the creation of a special district when a government-financed enhancement is made that benefits the residents of the special district. From that time on, two sets of tax records are maintained for the district: one that reflects the value of assets up to the time of the enhancement and one that reflects growth in assessed property value in the district after the enhancement. Tax revenues collected on the increased property values can be diverted to pay for the cost of the government-financed program in the special district. In some cases, governments issue tax-increment bonds for revitalization projects, with the bond being backed, in part, by the anticipated increase in property values resulting from the investment.

Tax-increment financing differs from a special assessment district in that property tax rates are increased in a special assessment district to cover improvements made in the district. In special districts utilizing tax-increment financing, tax rates may not be increased, but additional revenues are collected based on increased assessed property values enjoyed after the improvements are made.

REFERENCES

Office of Management and Budget. 1994. Catalog of Federal Domestic Assistance. Library of Congress No. 73–600118. Washington, D.C.: Government Printing Office.

Index